LIQUID CRYSTALS

Design, Behavior and Substituent Effects of 4-Ring Bent-Core Mesogens

ABOUT THE AUTHOR

Dr. Rahul Kanti Nath (b. 20 September 1981) is an accomplished academic and researcher in the field of Chemistry. He obtained his M.Sc. in Chemistry from Assam University, Silchar, in 2004 and went on to earn his Ph.D. from the same university in 2011, specializing in the synthesis and study of novel liquid crystalline materials.

With a keen interest in molecular design and materials chemistry, Dr. Nath has published six research papers in reputed international journals and contributed two chapters to edited academic books. His research focuses on liquid crystals, especially bent-core mesogens, and their structural and functional properties.

Since May 2017, Dr. Nath has been serving as an Assistant Professor in the Department of Chemistry at Cachar College, Silchar, where he is actively engaged in teaching, mentoring, and scientific exploration. His work reflects a strong commitment to academic excellence and the advancement of chemical science in both research and education.

LIQUID CRYSTALS
Design, Behavior and Substituent Effects of 4-Ring Bent-Core Mesogens

Dr Rahul Kanti Nath

ACADEMIC ENCLAVE

First Edition, 2025

ISBN: 978-93-48642-94-3

Published by
ACADEMIC ENCLAVE™
House No. 10/15, Ground Floor,
Block-A, Street No. 14/3, Kamal Vihar,
Kamal Pur, Burari, Delhi-110084
Email: info@academicenclave.com
Mobile: +91-7042423166
Website: www.academicenclave.com

Printed at Replika Press Pvt. Ltd.

In loving memory of

Rupantar

"Forever in my heart"

Preface

Liquid crystals continue to captivate scientific inquiry owing to their remarkable combination of fluidity and long-range molecular order. Among the diverse classes of liquid crystalline materials, bent-core or "banana-shaped" mesogens have emerged as a particularly intriguing category, offering unique structural motifs and exhibiting exotic phase behaviors not typically observed in conventional calamitic systems.

This book presents an in-depth exploration into the design, synthesis, and characterization of novel achiral bent-core compounds based on a four-ring molecular architecture. Central to this study is an investigation into how subtle molecular modifications, particularly the nature and position of substituents can dramatically influence the mesomorphic and emissive properties of these materials. The findings not only contribute to a deeper understanding of the structure–property relationships in bent-core liquid crystals but also offer valuable insights for the rational design of next-generation soft materials with tunable functionalities.

The research draws upon a wide array of experimental techniques, including differential scanning calorimetry (DSC), polarized optical microscopy (POM), and X-ray diffraction (XRD), to study phase transitions and structural organization. In parallel, the synthetic strategies employed highlight efficient pathways to achieving target molecular designs with precise functional tuning. Emphasis is placed on both polar and non-polar substitution patterns at key positions within the molecular core, revealing their significant effects on phase stability, emergence of chiral behavior, and photophysical characteristics.

Through this work, an attempt has been made to bridge molecular-level design principles with macroscopic phase behavior, thereby contributing to the broader field of liquid crystal science and materials chemistry. It is my hope that the insights presented here will be of interest to researchers working in the domains of soft matter, molecular self-assembly, and organic functional materials.

Dr Rahul Kanti Nath

Contents

CHAPTER 1

Introduction Liquid Crystals

This chapter deals with the general attributes of different liquid crystalline systems along with the brief introduction about these materials, including the brief history of the liquid crystal. We have also given a details description about the different mesogenic phases identified in bent-core compounds, how they classified and their phase structures determined by X-ray diffraction and other techniques are described.

1. Introduction

The discovery of spontaneous polarity [1] and chirality [2] in smectic liquid crystals (LCs) of bent shaped molecules has stimulated a new research field of soft materials [3-7] attracting both chemists and physicists alike because of their distinct characteristics viz., spontaneous supramolecular chirality, ferro- and anti-ferroelectricity,electro-optic switching and exotic phases. The crucial prerequisite for achieving polar order and chirality in any of the materials is that the individual constituent molecule should possess a bent shape in the central core and form smectic layers in which the molecular planestilted, resulting in a spontaneously chiral architecture. The synthesis of Bent-core molecules reported for the first time at Halle University in 1929-32 [8, 9], was confirmed to be a banana phase, B6 or a nematic phase recently [10]. However at that time the LC phases of these bent-core molecules were not paid much attention. The majority of the bent-core or banana shaped compounds reported in literature contain five or more aromatic rings [3, 4]. Very few reports on bent-core four aromatic ring compounds exhibiting mesomorphism appeared in literature [11], while majority of them are either exhibiting calamitic type liquid crystalline phases or non- mesomorphic. Four-ring bent-core molecules can be classified as intermediate between the banana-shaped (five or six or seven ring systems) and hockey stick (i.e., calamitic mesogens with a alkyl end chain in the meta position of a calamitic molecule) shaped compounds. The design and synthesis of four-ring architectures exhibiting mesomorphism and further investigations on these compounds are at the boundary between calamitic and banana-shaped liquid crystals to make them a new field of liquid crystal research showing many new open questions. ***Hence it is a significant challenge to design and tune the four-ring bent-core molecules to obtain banana LC phases at the primary level and then to generate nematic phases at a distinct level.*** In this book a new class of four-ring unsymmetrical bent-core mesogens based on the use of simple ester linking group as the molecular bend have been described. In particular polarization modulated layer undulated B7/B1$_{RevTilted}$ phases and chiral nematic phases exhibited by these compounds are the high lights of the contemporary research.

1.1 Liquid Crystal

The liquid crystalline state may be identified as a distinct and unique fourth state of matter [12-14] which is characterised by the dual properties which of solid and liquid (Figure 1.1).The intermediate thermodynamically stable phase i.e., liquid crystalline state possesses the dual characteristics viz., anisotropic physical properties of solid and fluid properties of liquid. The salient features of this characteristic fourth state of matter are fluid properties of liquid viz., flow, viscosity, surface tension etc. and anisotropic properties of a crystal viz., optical, electrical and magnetic anisotropy etc. The liquid crystalline phases possess orientational and in some phases additionally positional order. It was first recognised in the middle of the 19th century through the study of nerve myelin and derivatives of cholesterol. The research in this area gathered momentum during 1960-70 with the invention of transparent conductive electrodes and room temperature liquid crystals. Further the pioneering work of Gray in early 1970's in synthesis of cyanobiphenyls derivatives exhibiting liquid crystal behaviour at ambient temperatures, which are being useful in electrooptical devices, had motivated several researchers across the globe in this area.

a b c

Figure 1.1: Molecular arrangements in (a) crystals,(b) liquid crystals and (c) liquids.

A compound which displays liquid crystal properties is referred to as a *mesogen* and said to exhibit *mesomorphism.* Liquid crystals may be considered either as disordered solids or ordered liquids, and their properties are very dependent on temperature and the presence or absence of solvent.

1.2 History of Liquid Crystals

History shows that lyotropic liquid crystals were discovered 30 years before omit than thermotropic liquid crystals. In the early 1850's, Virchow (1821-1902) a physician and pathologist described a soft, floating substance from nerve core. He named it "Myelin". In 1857, Matteneheimer (1824-1898) discovered that Myelin was birefringent. The foundation of liquid crystal science is traditionally set in the year 1888, with the work of Reinitzer. Reinitzer was a botanist, although in modern terms he would perhaps be thought as a biochemist. He was at that time 30 years old and an assistant to Professor Weiss at the Institute of Plant Physiology at the German University of Prague. Reinitzer's experiments involved extracting cholesterol from carrots in order to determine its chemical formula, which at that time was unknown. Reinitzer found, to his amazement, that this compound does not melt like other compounds. Cholesteryl benzoate exhibited *two* melting points. At 145.5 °C the solid melted into a cloudy liquid. The cloudy liquid lasted up to 178.5 °C, at which point the cloudy liquid suddenly became clear. Furthermore, the phenomenon appeared to be reversible. Near both transition points the system exhibited some dramatic colours. Both the colours and the double melting were worthy of note. Reinitzer sought the help of Lehmann, a well-known crystallographer, such that his expertise on polarising microscope might clarify the situation. There followed an exchange of letters, and presumably of samples as well, throughout March and April of 1888. Lehmann examined the intermediate cloudy fluid, and reported that he had seen crystallites. When the exchange of letters ended on April 24, although definitive answers to the nature of the cloudy phase had not been elicited, Reinitzer felt that he had enough to publish. The important point here is that these first observations of liquid crystals were a serendipitous by-product of an apparently unrelated and evidence lacking piece of research. Neither for the first nor for the last time, nature had a sprung a surprise on an unprepared investigator. Table 1.1 depicts some of the molecular structures of important discoveries in the area of liquid crystals.

Table 1.1: Important liquid crystal discoveries

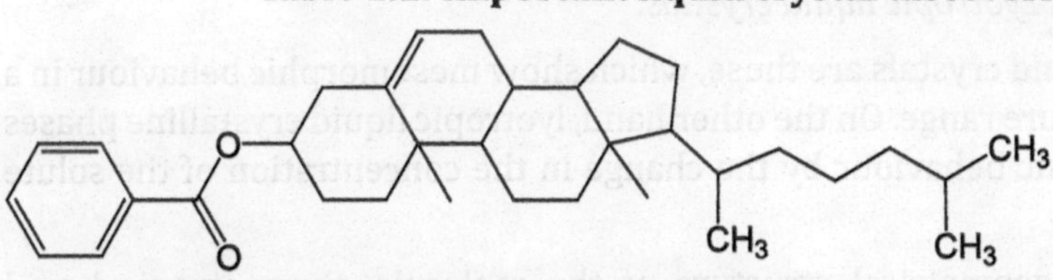

FIRST THERMOTROPIC LIQUID CRYSTAL

Common abbreviation: Cholesterol Benzoate (CB), [1888].Cr 150.5 N* 182.6 Iso.

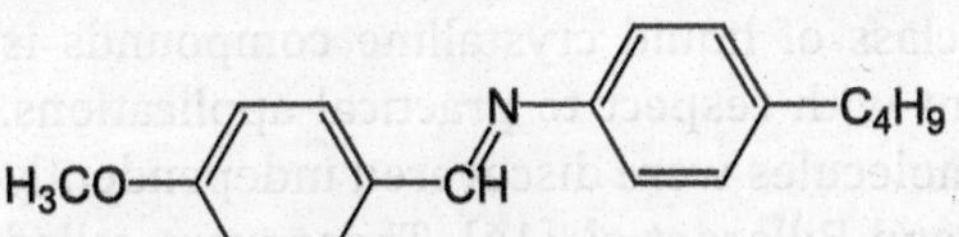

FIRST ROOM TEMPERATURE NEMATIC LIQUID CRYSTAL

Common abbreviation: MBBA, [1969], Cr. 22.0 N 48.0 Iso.

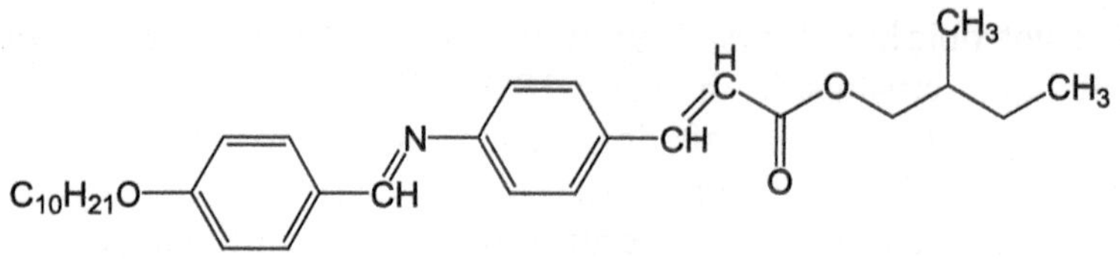

FIRST FERROELECTRIC LIQUID CRYSTAL

Common abbreviation: DOBAMBC, [1977],Cr. 76.0 C* 95 A 117.0 Iso.

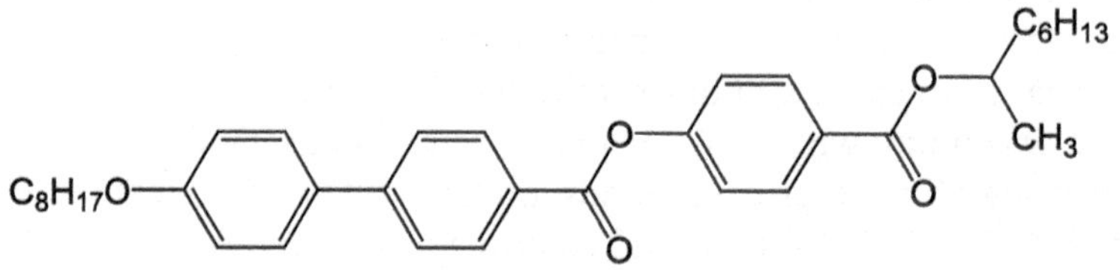

FIRST ANTI-FERROELECTRIC LIQUID CRYSTAL,

Abbreviation: MHPOBC, [1989], Cr. 84.0 CA* 18.4 Cγ 119.2 C* 120.9 Cα 122.0 A 148.0 Iso.

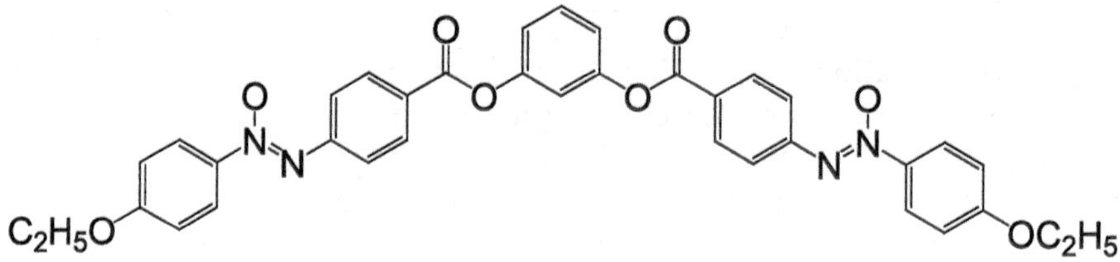

FIRST BANANA SHAPED LIQUID CRYSTAL,

Abbreviation: NOBOW, [1996], Cr 186 B_6 224 Iso.

1.3 Classification of Liquid Crystals

Liquid crystals can be divided from the chemical standpoint into: *Thermotropic liquid crystals* and *Lyotropic liquid crystals.*

Thermotropic liquid crystals are those, which show mesomorphic behaviour in a definite temperature range. On the other hand, lyotropic liquid crystalline phases show mesomorphic behaviour by the change in the concentration of the solute in a solution.

On the basis of geometrical structure of the molecule, thermotropic liquid crystals can be divided into rod like, bow like and disc like. The "classical" liquid crystals are derived from rod like molecules called "CALAMITIC" (derived from the Greek word *colomos*= rod). This class of liquid crystalline compounds is highly investigated and most important with respect to practical applications. Liquid crystals derived from disc like molecules were discovered independently by Chandrasekhar *et al.* [15] in 1977 and Billard*et al.* [16]. These were called *DISCOTIC* liquid crystals. The synthesis of Bent-core molecules reported for the first time at Halle University during 1929-32 [8, 9] and were not identified then. These classes of compounds are also called "BANANA" or "Bent-core" or "V"-shaped liquid crystals.

1.3.1 Calamitic liquid crystals

However, a vast majority of thermotropic are composed of rod-like molecules. They exhibit a variety of phases with the variation in temperature. It was in 1922, Friedel [17-19], a French scientist proposed a classification scheme based on the molecular structural arrangement where he named different liquid crystal systems: nematic, smectic and cholesteric.

Nematic liquid crystals

Nematic (N) is the most common and simplest liquid crystal phase. The word "nematic" is derived from the Greek word for *"Thread"* as thread like textures were observed under the polarizing microscope. In this phase, the molecules maintain a preferred orientational direction as they diffuse throughout the sample, but there is no translational order as shown in Figure 1.2.

Smectic Liquid Crystals

The word "smectic" is derived from the Greek word for *"soap"*. This seemingly ambiguous origin is explained by the fact that the thick, slippery substance often found at the bottom of a soap dish is actually a type of smectic liquid crystal [21]. Molecules in this phase show additional degrees of translational order not present in the nematic phase. In the smectic phase, the molecules maintain the general orientational order of nematics, but also tend to align themselves in layers; with a well-defined interlayer spacing, which can be measured by X-ray diffraction [22].

a b c

Figure 1.2: Arrangements of molecules in (a) nematic, (b) smectic A and (c) smectic C phase.

Table 1.2: Structural Classification of Smectic Liquid Crystals

Smectic A:	Liquid-like layers with molecules upright on the average. Negligible in-plane and interlayer correlations. Thus, this phase can be classified as an orientationally ordered liquid on which a 1-D density wave is superimposed.
Smectic B:	Basically similar to Sm A phase but with higher in-plane ordering. Two distinct types of smectic had been identified, (a) Crystal B, 3-D crystal having hexagonal lattice and weak interlayer ordering. (b) Hexatic B, stack of interacting hexatic layers with short-range in plane ordering and 3-D six-fold bond orientational ordering.
Smectic C:	Liquid-like layers as in Sm A, but with the molecules inclined with respect to layer normal.
Smectic C*:	Chiral Smectic C with twist axis normal to the layers.
Smectic D:	Cubic lattice with micelle type molecular arrangement. This phase is usually observed between Sm A and Sm C or between Sm C and Isotropic phase.
Smectic E:	3-D orthorhombic crystal with interlayer herringbone structure.
Smectic F:	Titled hexatic structure with C centered monoclinic (a>b) with in-plane short-range positional correlation and weak or no interlayer positional correlation.
Smectic G:	3-D crystal with C-centered monoclinic (a>b).
Smectic H:	3-D crystal with C-centered monoclinic (a>b) and having herringbone structure.
Smectic I:	Titled hexatic C-centered monoclinic (b>a) with slightly greater in plane correlation than Sm F.
Smectic J:	3-D crystal with C centered mono clinic (b>a) structure.
Smectic K:	3-D monoclinic (b>a) structure with herringbone structure.

Cholesteric Liquid Crystals

When a nematic phase is incorporated with a chiral molecule then the resulting structure can be visualized as a stack of very thin 2-D nematic layers with the director in each layer twisted with respect to those above and below. This induces a helical director configuration in which the director rotates through the material. Hence, the name twisted nematic (N_t) is assigned to this class of liquid crystals. The molecules first to form this phase were derivatives of cholesterol, so they are also called cholesteric liquid crystal. An important characteristic of the cholesteric mesophase is its pitch. The pitch, p, is defined as the distance it takes for the director to rotate one full turn in the helix.

A B

Figure 1.3: (a) molecular arrangement in cholesteric phase (b) pitch of cholesteric phase.

1.3.2 Discotic liquid crystals

In 1977 Chandrasekhar et al [15] demonstrated that disk shaped molecules such as hexa-n-alkanoyloxybenzenes exhibit mesomorphism. Based on optical, calorimetric and X-ray studies they proposed a structure for the this new mesophase in which the molecules are staked one on top of the other to form columns [13], Which in turn arrange themselves into two-dimensional lattices as shown below in Figure 2. Discotic systems in general exhibit nematic (N) and/or columnar (Col) phases.

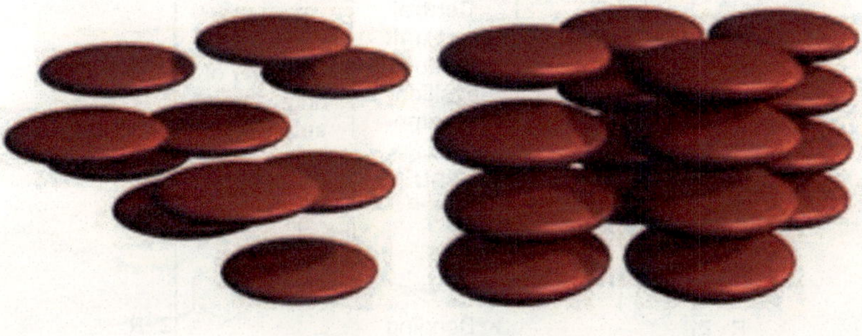

a b

Figure 1.4: Molecular arrangement of (a) discotic nematic and (b) columnar mesophase.

1.3.3 Banana Shaped liquid crystals

Bent-core or banana-shaped liquid crystals refer to the liquid crystals whose molecular shapes resemble the bananas or a bow with a bend at the centre of the molecules and also exhibiting the optical textures similar to banana leaf like. Majority of the bent-core molecules link through an aromatic 1,3-disubstituted phenyl ring (Figure 1.6)or 2,7-disubstituted naphthalene moiety derivatives. The angle between two linking arms ranges from 100° to 140°. The first bent-core liquid crystal was synthesized in the research group of Vorlander by Schroder in 1923 [20], but the mesophase properties of the compound were not reported. Even though in 1932, the observation of mesophases in bent-core compounds appeared in literature for the first time [8], the identified nomenclature of banana phases was unknown at that time.

Bent-core or banana-shaped molecules exhibiting mesomorphism did not get too much attention until Matsunaga [24, 25] synthesized new mesogenic compounds with "banana-shaped" molecular structures. In 1996 in 6th ILCC at Kent, Takezoe presented the first mesophase with antiferroelectricity formed by achiral bent-core molecules.This phase was later known as the B_2 phase (tilted polar smectic phase). This discovery is the beginning of interest in these new mesophases which are distinctly different in physical properties fromthat of smectic phases exhibited by calamitic (rod-like) molecules except layer ordering. Hence most of the research efforts hadbeen focused on the bent-core smectics, because nematic phases were meagre or rather uncommon in different homologous series of compounds reported in recent years. Till now a great number of bent-core mesophases have been predicted theoretically, or reported from experiments. Among them there are some classical phases such as smectic A, smectic C and the new ones actually classified with names ranging from B1 to B8[7, 25] defined at the 1997 Berlin workshop[26] and after. The B1 to B8 series represent phases in the sequence of their discovery.

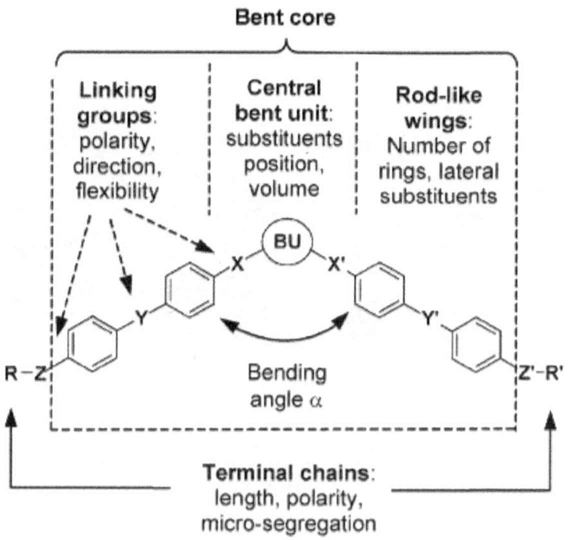

Figure 1.5: General structure of a bend core molecule

1.4 Liquid Crystal Phases of Bent-core Molecules

Bent-core molecules exhibit several meso-phases, some of which are significantly different from the liquid crystal phases of rod-like molecules. Because of the characteristic molecular shape and based on symmetry properties bent-core mesogens produce unique smectic phases that rod-like molecules never formed. After the Berlin workshop on 1997 [26], liquid crystal phases of bent-core molecules have been named as the 'B*i*' (*i*is an integer number) phase in the chorological order of their discovery.The B1 phase exhibits a two-dimensional order. There are strong arguments that it is necessary to subdivide phase B1 into modifications with different columnar structures[27]. The B2 phase corresponds to the phase first discussed by Nioriet.el. [1] is of a special interest because it is the first liquid crystal showing antiferroelectric behaviour. Another phase which shows similar electro-optical switching properties was named as B5[28]. B3 and B4 are crystalline-like modifications, while the B4 phase is designated as the "blue phase" of bent-core mesophases[29, 30]. The B6 phase can be compared with an interdigitated SmC phase generated by bent-core molecules [17]. The helical filaments along with myelin, ribbon and spiral like optical textures are the characteristics of B7 phase[31, 32].

In the Boulder workshop on 2002, the nomenclature of these liquid crystal phases were discussed and people agreed to make another rule of naming. The most absolute definition of the phase would be using the space group of the phase but it was not convenient for all the people. For both convenience and precise, people almost agree to name the phases of bent-core mesogens using the same name used for the rod-like mesogens, for example Sm-*A* or Sm-*C*, and additional characters and suffixes. Also the space group of the phase will be added for a detailed definition.

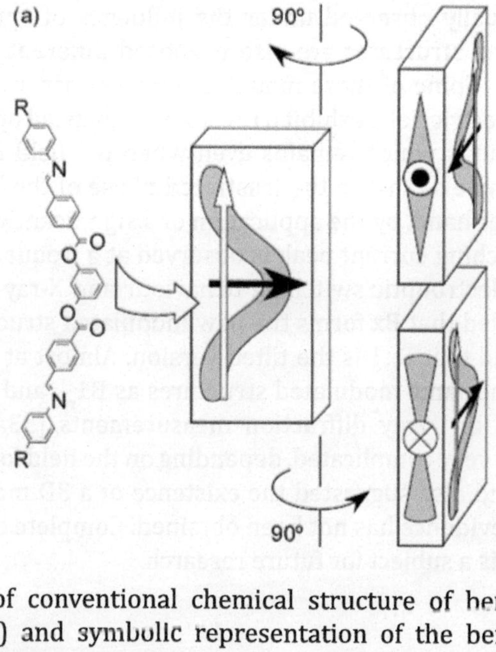

Figure 1.6: One of conventional chemical structure of bent-core molecules P-n-(O)-PIMB (left) and symbolic representation of the bent-core molecules throughout this book (right). White arrow in right figure represents molecular long axis, n, black arrows represent bent-director b.

1.4.1 B1 Phase

The B1 phase[Figure 1.7] exhibits two periodicities like a columnar phase, i.e. smectic layer ordering and another periodicity in a plane of smectic layers. Three different structures have been reported[7, 33] as shown in figure1.8(c-e).

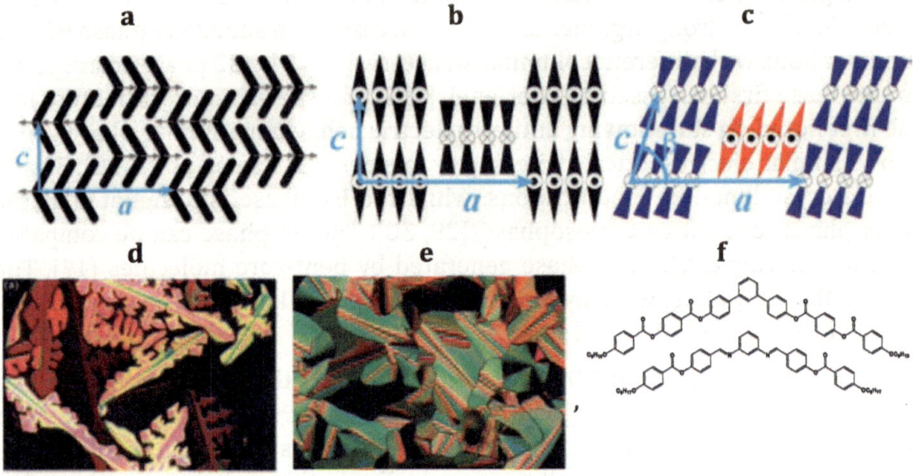

Figure 1.7: (a,b,c) Three proposed structures of the B1 phase viz. B1, B1$_{rev}$ and B1$_{revTilted}$, (d-e) Photomicrograph of the B$_1$ phases, (f) molecular structure of the compound showing B$_1$ phases.

All of these phases do not exhibit macroscopic spontaneous polarization and no response is usually observed under the influence of applied electric field. Different modulated structures are also proposed different from conventional B1 phase [27, 34]. Some of these modulated phases are switchable under the application of an electric field, exhibit irreversible enhanced optical birefringence in fan-shaped textures which remains even when the field is turned off. This behaviour is the same as that in the frustrated phase of the bent-shaped dimer [33b]. On the other hand, by the application of a significantly high field (>15 V/mm), a single switching current peak is observed at a frequency as low as 1 Hz. Considering this electrooptic switching behaviour and X-ray study, Szydlowska et al. [33c] concluded that Bx forms the new modulated structure (B1$_{rev}$) shown in Figure1.7(b), and that Bx1 is the tilted version. Almost at same time, Pelz et al. also reported the same modulated structures as B1$_{rev}$ and B1$_{revtilt}$ in different compounds based on X-ray diffraction measurements [33a]. The switching process in B1$_{revtilt}$ is very complicated, depending on the field form, amplitude, and frequency [33a] and also suggested the existence of a 3D modulated structure, although decisive evidence has not been obtained. Complete classification of the B1-related phases is a subject for future research.

1.4.2 B2 phase

The B2 phase is also known as the Sm-*CP* phase, in which molecules forming a smectic layer are tilted with respect to the layer normal (Sm-*C*) and the smectic layer exhibits macroscopic spontaneous polarization perpendicular to the molecular tilt plane. Now the word "B2" is used as a general term for the phase of simple Sm-*CP* structure in which four Sm-*CP* variations exist; Sm-C_SP_A, Sm-C_AP_A, Sm-C_SP_F and Sm-C_AP_F [figure 1.9]. In two of them, Sm-C_SP_A and Sm-C_SP_F, molecules in adjacent smectic layers tilt in the same direction, and molecules alternate the tilting direction in the Sm-C_AP_A and Sm-C_AP_F structure. Also, two of these structure is ferroelectric (Sm- C_SP_F and Sm-C_AP_F) and the other two are antiferroelectric (Sm-C_SP_A and Sm-C_AP_A). In terms of layer chirality, two of these structures, Sm-C_AP_A and Sm-C_SP_F are composed of layers of the same chirality (homogeneously-chiral or uni-chiral) and the other two, Sm-C_SP_A and Sm-C_AP_A are composed by stacking of two layers of opposite handedness (racemic, anti-chiral or hetero-chiral).

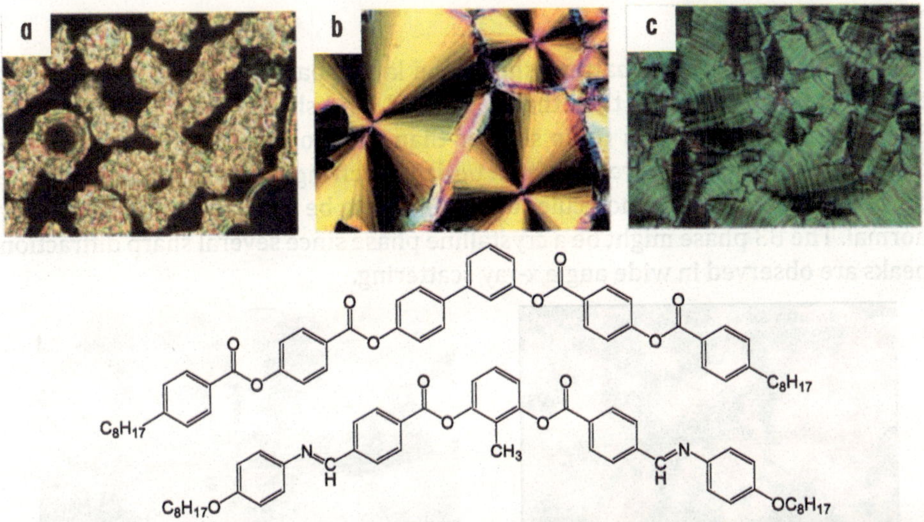

Figure 1.8: (a)Growth of the B2 phase on cooling from the isotropic liquid (b) fan shaped and (c) focal conic texture with stripes for B2 phase, (d) molecular structure of some compounds showing B2 phase.

In the B2 phase of achiral bent-core molecules, two antiferroelectric structures are often observed as the ground state structure. Ferroelectric B2 phase is much rare to be observed [2,35-40].

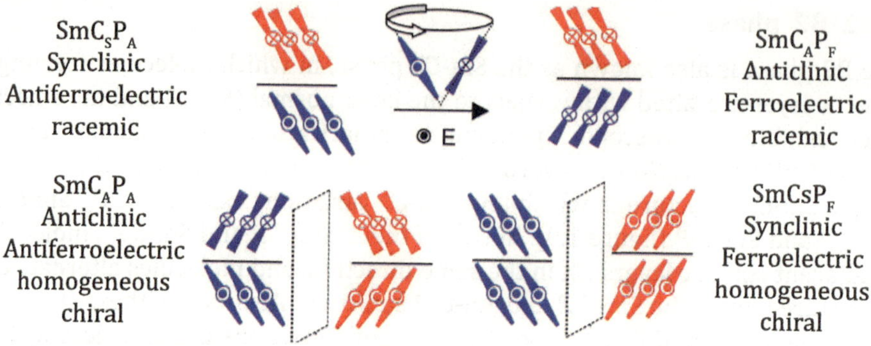

Figure. 1.9: The supramolecular arrangements resulting from the combination of the different tilt directions and polar directions (abbreviations: Cs = synclinic tilt, Ca = anticlinic tilt, P_F = FE polar order, PA = A_F polar order).

1.4.3 B3 Phase

The 2D X-ray diffraction pattern shows the layer spacing in B3 is the same as that in the crystal, longer than that in B2 [34] and shorter than that in B4 [2]. Since the layer spacing in B3 is shorter than the molecular length of alltrans-conformation and a texture with high optical birefringence was observed in the homeotropic alignment, molecules are expected to be tilted away from the layer normal. The B3 phase might be a crystalline phase since several sharp diffraction peaks are observed in wide angle x-ray scattering.

Figure. 1.10: Photomicrographs of the B_3 phase

1.4.4 B4 Phase

The first observation of spontaneous formation of macroscopic chiral domains was made in the B4 phases for the prototype bent-core molecules (Figure 1.11). According to X-ray diffraction this phase is characterised by a series of rather sharp, but not resolution limited reflections in the wide-angle range and hence, it seems to be a highly ordered, probably crystalline mesophase. The molecules are non-tilted and the B4 phases cannot be switched under electric fields. An additional characteristic textural feature of this mesophase is a visible blue colour, the origin of which is not clear. A helical structure [40] composed of slabs with a helix axis parallel to the layers, similar to twist grain boundary phases (TGB-like)[41, 42] had been proposed for this mesophase. The fact that the molecules are non-tilted excludes the possibility of super-structural (layer) chirality, which would require a tilted organisation. Hence, in this mesophase chirality can originate either from the TGB helix itself, or from a conformational molecular chirality.

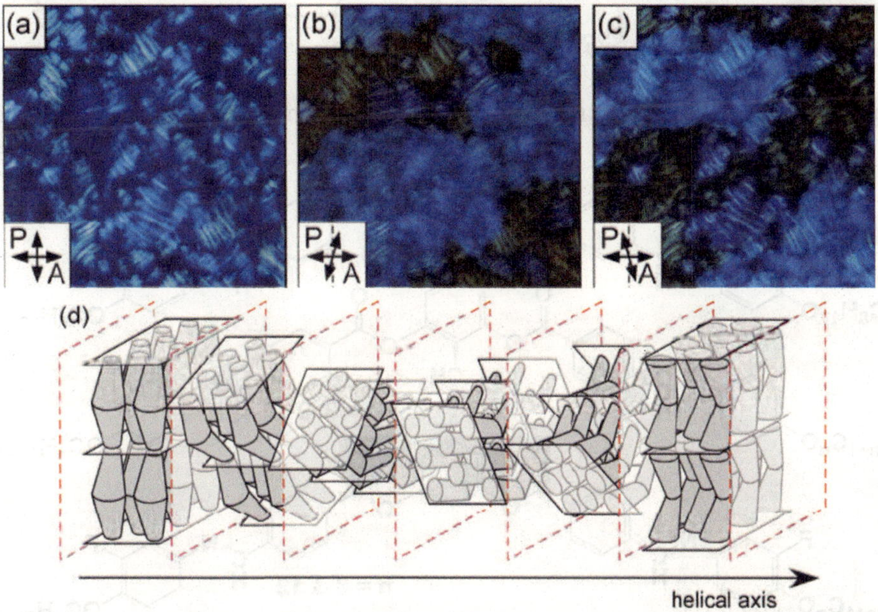

Figure 1.11: Optical texture of B4 phase a)crossed polarisers b) polarizer rotated clockwise and c) anticlockwise. The molecular arrangement in twisted grain boundary (TGB) phases (B4).

1.4.5 B5 phase:

The B5 phase is the Sm-*CP* phase with the in-plane positional ordering. Electro-optic response in this phase is similar to that in the B2 phase[28, 43]. From x-ray scattering measurements, the phase is known to have a positional ordering in the smectic layer. Weissfloget *al.* reported three distinct B5 phases in which electro-optic response and x-ray scattering profile were the same but distinguishable only in DSC measurements [44].

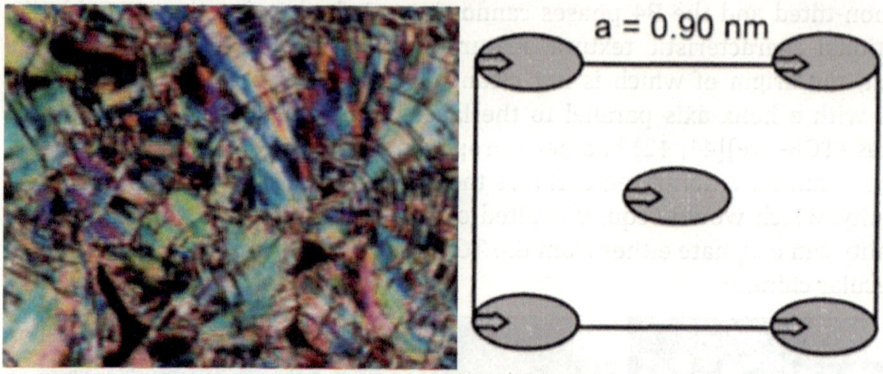

Figure 1.12 a: Optical textures and arrangement of molecules in the B5 phase.

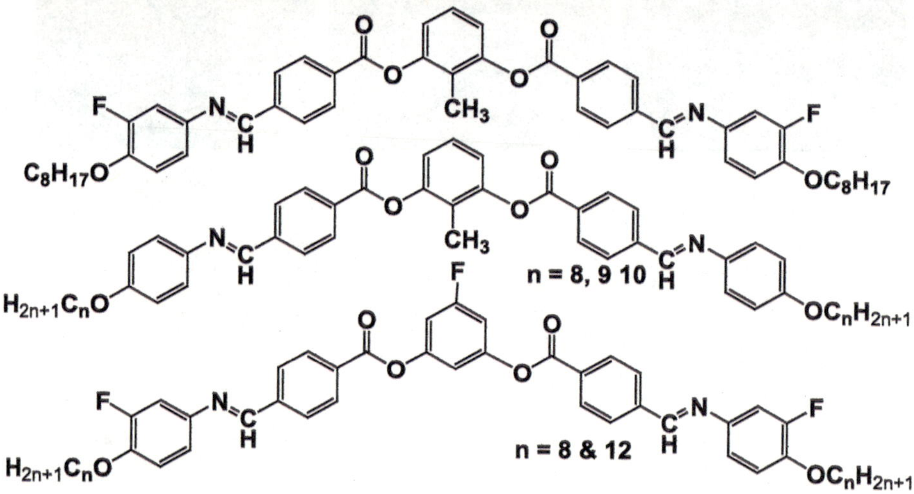

Figure 1.12b: Molecular structures of some compounds exhibiting B5 phase

1.4.6 B6 phase:

The B6 phase [figure 1.13] is similar to the B1 phase in terms of interdigitation of molecules each other for the half molecular length, but this phase does not exhibit 2-dimensional ordering which easily determined by x-ray scattering experiments. Also the phase exhibits a reflection corresponding to the half molecular length (002) while the reflection corresponding to the molecular length (001) is cancelled out and not observed.The B6 phase is sometimes observed in an upper temperature range of the B1 phase [37]. Because dipole moments of molecules are cancelled out each other, no electro-optic response is observed except a dielectric response.

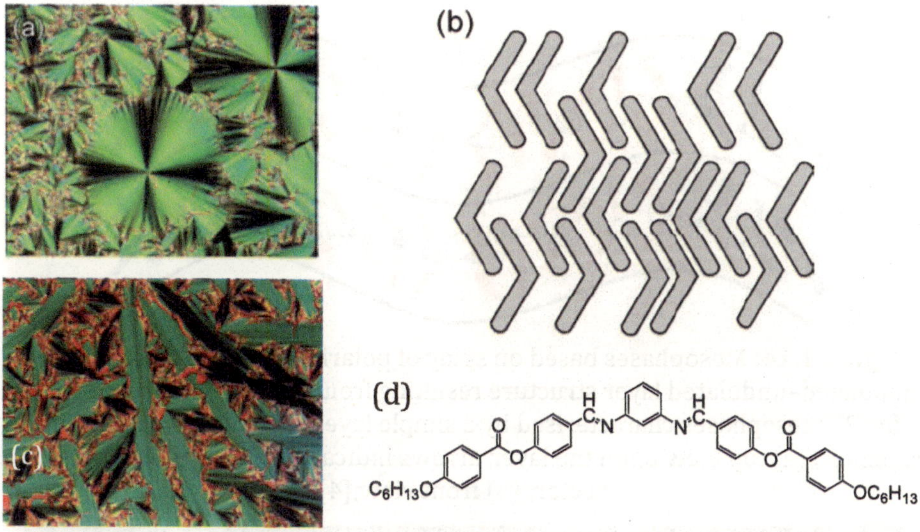

Figure 1.13: (a,c) Different POM textures of B6 phase, (b) the model for B6 phase, (d) molecular structure of the compound showing B6 phase.

1.4.7 B7 phase

The B7 phase is one of the important constituent of the Sm-*CP* family [45]. The detailed molecular structure of B7 phase illustrated in Figure 1.15 and was recently investigated by means of X-ray scattering and observations of transmission electron microscope micrographs of the quenched specimen of a thin film of the sample generated by freeze fracture technique [46]. The B7 phase was first defined by the texture observations using optical microscope, which exhibited huge variation of spiral domains, helical ribbon like domains, focal conics and leaf-like domains [47]. Textures of the B7 phase are shown in figure 1.15. From high-resolution x-ray measurements, several different structures that show the same texture under an optical microscope are known to exist. Hence the name "B7" is often used as the name of texture of phase and the phase is called as the *polarization modulated layer undulated* ***(PMLU)***phase.

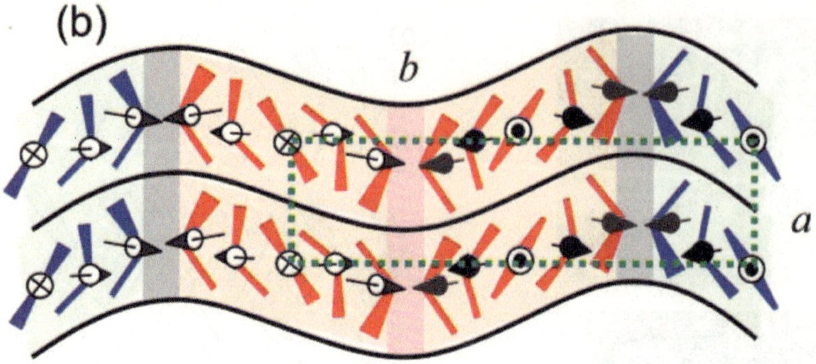

Figure 1.14: Mesophases based on splay of polarisation; (a, b) model of the modulated–undulated layer structure resulting from splay defects, as suggested for B7 mesophases characterised by a simple layer structure and additional satellites; (a) top view upon the layer, arrows indicate the direction of the polar vector; (b) front view;[4]

Figure 1.15: Typical textures observed in B7 phase, (a), (b) In a 4-micron thick cell, (c) Racemic MHO-Bow in a 6-micron thick cell and (d)-(f) Chiral MHO-Bow in a 8-micron thick cell.

1.4.8 B8 phase

This phase was discovered recently in a symmetric bent compound as shown in Figure 1.16 [48]. The homologues of this compound with shorter end chains did not exhibit the B8 phase. The microphotographs of the B8 phase are displayed in Figure 1.17 a, b and c. Upon gradual cooling from the isotropic phase, spiral domains grow as in the B7 phase. By further cooling, a fan-shaped texture is also observed, in which the extinction direction is parallel to the layer normal, and antiferroelectric-type switching is observed.TheX-ray diffraction suggests that the B8 phase has a bilayer structure. This is the first and only report on the B8 phase thus far, and further investigations are expected.

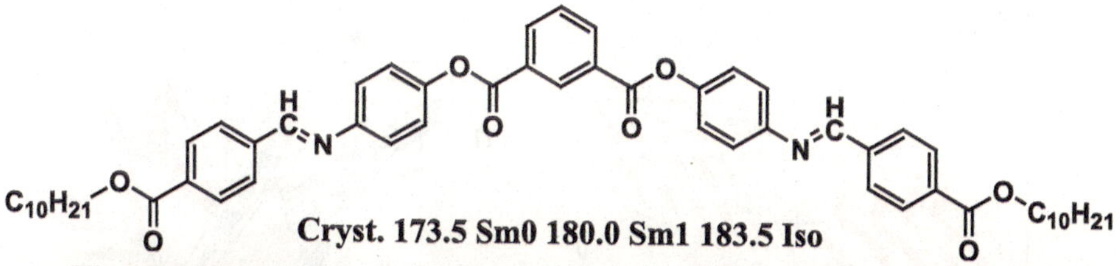

Figure 1.16: Chemical structures of compounds exhibiting B8 phase

Figure 1.17: Microphotographs of texture observed in B8 phase at (a) 181°C, (b) 175°C and (c) 166°C.

The banana phase structures and properties of smectic phases of bent-core molecules are summarized on Table 1.4

Table 1.4: Banana phase structures and properties of bent-core molecules

Phase	Structure	Polar	Tilting in layer
B1	Frustration (columnar)	Usually non-polar	Tilted, non-tilted
B2	SmCP	polar	Tilted
B3	Sm (crystal like)	non- polar	non-tilted
B4	TGB (chiral domain)	non- polar	non-tilted
B5	SmCP (hexatic order)	Polar	Tilted
B6	Interdigitation	non- polar	Tilted, non-tilted
B7	SmCP (modulated)	Polar	Tilted
B8	SmCP	Polar	Tilted

1.5 Chirality in Achiral Banana Shaped Molecule

In the tilted polar smectic phases (Sm-*CP* phases) of *achiral* bent-core molecules, two types of layers exist, and one of which is the mirror image of the other [figure 1.18 (a)]. As in the case of the Sm-*C** phase of rod-like molecules, the spontaneous polarization(P) of each layers is uniquely defined. However, the situation is little complicated because molecules in each layer have a bent-direction b. P and b are assumed to be parallel but their directions are not always the same. For this case, the layer chirality, instead of the sign of spontaneous polarization, is defined by Link *et al*[29] using the molecular tilt direction and b.

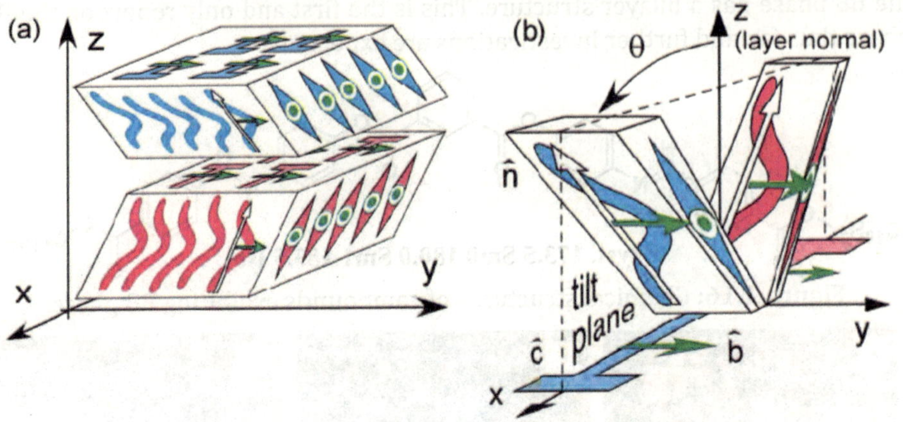

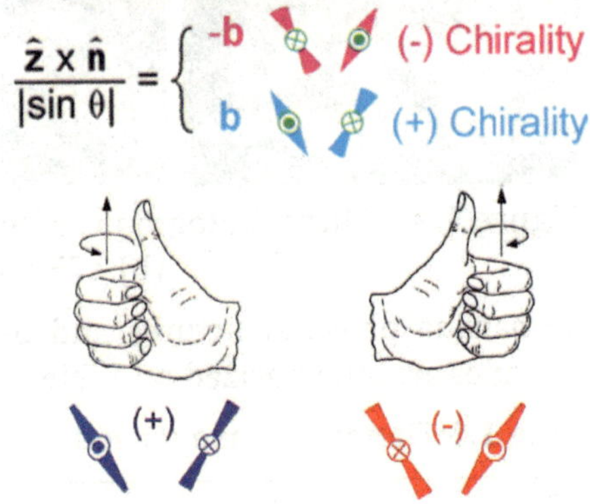

Figure 1.18: (a) Geometry for smectic layers and layer stacking in tilted smectic phase of bent-core molecules. (b) In one smectic layer with layer normal along the z-axis composed of bent-core molecules with polar order along b parallel to the y-axis, two possible layer structures one of which is mirror image of the other, are defined by the director n tilting by θ either along the positive or negative to x-axis.

Banana molecules represent the first achiral molecules forming super-structural chirality in the fluid liquid crystalline state [15]. It arises due to the tilted organisation of the molecules within the polar layers. In this configuration the layer normal, the tilt direction and the polar axis define a chiral system which is either right-handed or left-handed as shown in Figure. 1.18. Changing either polarisation direction (indicated by crosses or dots in circles) or tilt direction changes the chirality sense of the layer (indicated by red (+) vs. blue (-) colour) [2]. There are four possible combinations of the layers. In SmC_AP_A and SmC_SP_F structures, all the layers of a macroscopic domain have the same chirality sense. These phase structures are homogeneous chiral and occur in both enantiomeric forms (as conglomerates). In the SmC_SP_A and SmC_AP_F structures there are equal numbers of layers with (+) and (-) chirality and hence these superstructures are racemic on a macroscopic scale [2]. During the switching process, rotation usually takes place on a cone, which changes tilt direction and polar direction simultaneously and therefore chirality is maintained. This means that switching under electric fields occurs either between the two chiral states SmC_AP_A and

SmC_SP_F or between the two racemic states SmC_SP_A and SmC_AP_F[50, 51] (figure 1.19).

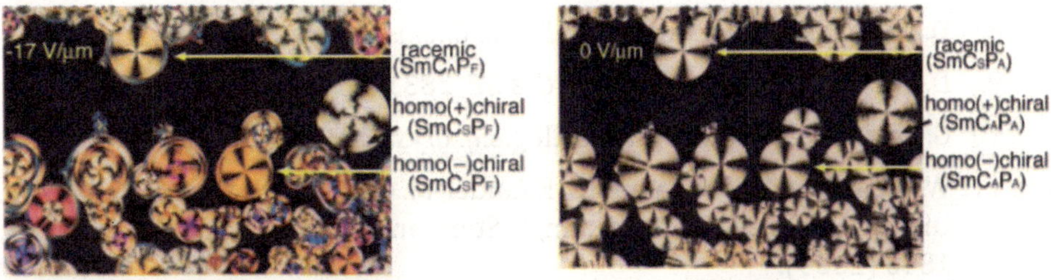

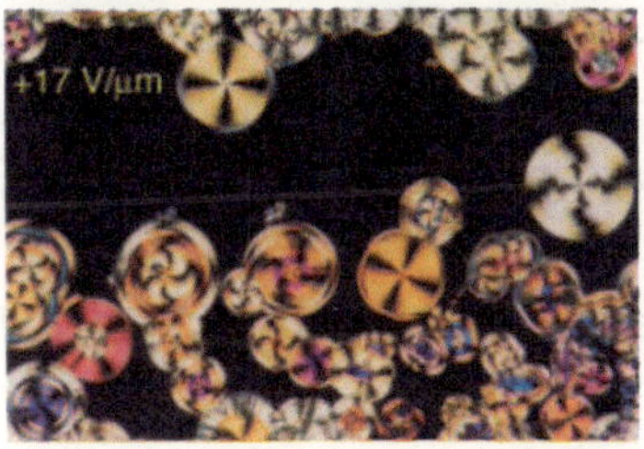

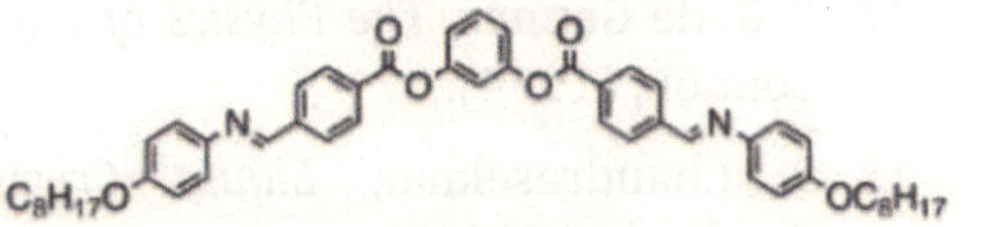

Figure 1.19: Microphotographs in B2 phase of P-8-O-PIMB. A variety of colored domains are observed, which indicates that the homochiral, racemic, and partial mixing of homochiral and racemic domains coexist.

This type of chirality is observable during the switching process. Further the intrinsic super-structural chirality of these polar smectic and modulated smectic phases due to the combination of tilt and polar order, which is manifested in the switching process (rotation of extinction crosses) there is also a spontaneous desymmetrisation[45], observed as chiro-optical effect in some of these mesophases. These mesophases exhibit chirality characterised by the spontaneous formation of a conglomerate composed of macroscopic domains of opposite chirality. Between crossed polarisers the textures appear as nearly uniformly blue or dark areas. By slight decrossing of the polarisers, domains become visible and the brightness of the domains is exchanged by decrossing the polarisers i.e. rotation of polarizer either in clockwise or anticlockwise direction, indicating the optical activity (rotation of polarised light) of these domains. Different domains have opposite sense of chirality.

References

1. T. Niori, T. Sekine, J. Watanabe, T. Furukawa and H. Takezoe, *J. Mater. Chem.*, **6**, 1231 (1996).
2. D. R. Link, G. Natale, R. F. Shao, J. E. Maclennan, N. A. Clark, E. Korblova and D. M. Walba, *Science*, **278**, 1924 (1997).
3. H. Takezoe and Y. Takanishi, *Jpn. J. Appl. Phys.*, **45**, 597 (2006).
4. R. A. Reddy and C. Tschierske, *J. Mater. Chem.*, **16**, 907 (2006).
5. M. B. Ros, J. L. Serrano, M. R. de la Fuente and C. L. Folcia, *J. Mater. Chem.*, **15**, 5093 (2005).

6. C. Tschierske and G. Dantlgraber, *Pramana*, **61**, 455 (2003).
7. G. Pelzl, S. Diele and W. Weissflog, *Adv. Mater.*, **11**, 707 (1999).
8. D. Vorlander, and A. Apel.,*Ber. Dtsch. Chem.Ges.* **65**,1101 (1932).
9. D. Vorlander, *Ber. Dtsch. Chem.Ges.***62**, 2831 (1929).
10. G. Pelzl, I. Wirth and W. Weissflog,*Liq. Cryst.***28**,969 (2001).
11. W. Weissflog, U. Dunemann, S. F.Tandel, M. G. Tamba, H. Kresse, G. Pelzl, S. Diele, U. Baumeister, A. Eremin, S. Stern and R. Stannarius, *Soft Matter*,**5**, 1840 (2009).
12. P. G. de Gennes, *The Physics of Liquid Crystals*, Oxford University Press, London, (1974).
13. S. Chandrasekhar, *Liquid Crystals*, Cambridge University Press, Cambridge, (1992).
14. D. Demus, J. Goodby, G. W. Gray, H. W. Spiess and V. Vill,*Handbook of Liquid Crystals*, Vol. 1-3, ed. Wiley-VCH, Weinheim, (1998).
15. S. Chandrasekhar, B. K. Sadashiva and K. A. Suresh, *Pramana*,**9**, 471 (1977).
16. J. Billard, J. C. Dubois, T. H. Nguyen and A. Zann, *Nouveau J. Chim.***2**, 535 (1978).
17. G. Freidel, *Ann. Physique*,**18**, 273 (1922).
18. D. Demus and L. Richter, *Textures of Liquid Crystals*, Verlag Chemie, New York, (1978).
19. A.Cifferi, W. R. Krigbaum and R. B. Meyer, *Polymer Liquid Crystals*, Academic Press, New York, (1982).
20. D. Vorländer, *Z.Phys.Chem., Stoechiom. Verwandtschaftsl*, **105**, 211(1923).
21. P. J. Collings, *Liquid Crystals: Nature's Delicate Phase of Matter*, Princeton University Press, Princeton, (1990).
22. G. Vertogen and W. H. De Jeu, *Thermotropic Liquid Crystals, Fundamentals*,Springer-Verlag, Berlin, (1987).
23. Y. Matsunaga, S.Miyamoto, *Mol. Cryst. Liq. Cryst*, **237**, 311 (1993).
24. H. Matsuzaki, Y.Matsunaga, *Liq.Cryst.*, **14**, 105(1993).
25. K. Pelzl, W. Weissflogamd S. Diele, *6th European conference on Liquid Crystals*, Halle, Germany (2001).
26. Workshop on banana-shaped liquid crystals: Chirality by achiral molecules, Berlin, December (1997).
27. D. Kardas, M. Prehm, U. Baumeister, D. Pociecha, R. A. Reddy, G. H. Mehl and C. Tschierske: *J. Mater. Chem.*, **15**, 1722(2005).

28. S. Diele, S. Grande, H. Kruth, C. Lischka, G. Pelzl, W. Weissflog and I. Wirth,
29. *Ferroelectrics*, **212**, 169(1998).
30. H.T. Nguyen, J.C. Rouillon, J.P. Marcerou, J.P. Bedel, P. Barois, S. Sarmento, *Mol.Cryst. Liq. Cryst.*, **28**, 177 (1999).
31. A. Eremin, S. Diele, G. Pezl, H. N′adasi, W. Weissflog, J. Salfetnikova, H. Kresse, *Phys. Rev. E*, **64**, 051707 (2001).
32. Weissflog, H. Nadasi, U. Duneman, G. Pelzl, S. Diele, A. Eremin and H. Kresse, *J. Mater. Chem.*,**11**, 2748(2001).
33. R. A. Reddy and B. K. Sadashiva,*Liq. Cryst.*,**31**, 1069(2004).
34. (a) K. Pelz, W. Weissflog, U. Baumeister and S. Diele, *Liq. Cryst.*,**30**, 1151 (2003); (b) Y. Takanishi, T. Izumi, J. Watanabe, K. Ishikawa, H. Takezoe and A. Iida,*J. Mater. Chem.*,**9**, 2771(1999); (c) J. Szydlowska, J. Mieczkowski, J. Matraszek, D. W. Bruce, E. Gorecka, D. Pociecha and D. Guillon, *Phys. Rev.*E ,**67**, 031702 (2003).
35. (a) D. Shen, S. Diele, G. Pelzl, I. Wirth and C. Tschierske, *J. Mater. Chem.*, **9**, 661 (1999); (b) H. N. S. Murthy and B. K. Sadashiva,*J. Mater. Chem*, **15**,2056 (2005); (c)D. Shen, A. Pegenau, S. Diele, I. Wirth and C. Tschierske, *J. Am. Chem. Soc.*, **122**, 1593(2000); (d) R. A. Reddy, V. A. Raghunathan and B. K. Sadashiva, *Chem. Mater.*,**17**, 274(2005).
36. T. Sekine, T. Niori, M. Sone, J. Watanabe, S. Choi, Y. Takanishi and H. Takezoe, *Jpn. J. Appl. Phys.*, **36**, Part1, 6455 (1997).
37. G. Pelzl, S. Diele, S. Grande, A. Jakli, Ch. Lischka, H. Kresse, H. Schmalfuss, I. Wirth and W. Weissflog, *Liq. Cryst.*, **26**, 401 (1999).
38. W. Weissflog, Ch. Lischka, S. Diele, G. Pelzl and I. Wirth, *Mol. Cryst. Liq. Cryst.*, **328**, 101 (1999).
39. E. Gorecka, D. Pociecha, F. Araoka, D. R. Link, M. Nakata, J. Thisayukta, Y. Takanishi, K. Ishikawa, J. Watanabe and H. Takezoe, *Phys. Rev. E*, **62**, R4524(2000).
40. M. Nakata, D. R. Link, F. Araoka, J. Thisayukta, Y. Takanishi, K. Ishikawa, J. Watanabe, and H. Takezoe, *Liq. Cryst.*, **28**, 1301 (2001).
41. G. Dantlgraber, A. Eremin, S. Diele, A. Hauser, H. Kresse, G. Pelzl and C. Tschierske, *Angew. Chem. Int. Ed.*, **41**, 2408 (2002).
42. (a) J. W. Goodby, M. A. Waugh, S. M. Stein, E. Chin, R. Pindak and J. S. Patel, *Nature*, **337**, 449(1989);(b) S. R. Renn and T. C. Lubensky, *Phys. Rev.* A., **38**, 2132 (1988).
43. T. Sekine, T. Niori, J. Watanabe, T. Furukawa, S. W. Choi and H. Takezoe, *J. Mater. Chem.*,**7**, 1307(1997).

44. G. Pelzl, W. Weissflog*et al.*, Invited presentation I-2, 19th International Liquid Crystal Conference, Edinburgh, U.K. (2002)

45. G. Dantlgraber, D. Shen, S. Diele and C. Tschierske,*Chem. Mater.***14**, 1149(2002).

46. D. M. Walba, E. Körblova, R. Shao, J. E. Maclennan, D. R. Link, M. A. Glaser and N. A. Clark, *Science*, **288**, 2181 (2000).

47. D.A. Coleman, J. Fernsler, N. Chattham, M. Nakata, Y. Takanishi, D. R. Link, R. F. Shao, W.G. Jang, J.E. Maclennan, E. Körblova, O. Mondain, C. Boyer, W. Weissflog, G. Pelzl, L.C. Chien, D.M. Walba, J. Zasadzinski, J. Watanabe, H. Takezoe and N. A. Clark, *Science*, **301**, 1204 (2003).

48. G. Pelzl, S. Diele, A. Jakli, C. Lischka, I. Wirth, W. Weissflog, *Liq. Cryst.*, **26**, 135 (1999).

49. J. P. Bedel, J. C. Rouling, J. P. Marcerou, M. Laguerre, H. T. Nguyen and M. F. Achard, *Liq. Cryst.*, **28**, 1285 (2001).

50. M. Avalos, R. Babiano, P. Cintas, J. L. Jimenez and J. C. Palacios, *Tetrahedron: Asymmetry*, **15**, 3171 (2004).

51. M. Zennyoji, Y. Takanishi, K. Ishikawa, J. Thisayukta, J. Watanabe and H. Takezoe,*J. Mater. Chem.*,**9**, 2775(1999).

52. D. J. Earl, M. A. Osipov, H. Takezoe, Y. Takanishi and M. R. Wilson,*Phys. Rev.* E,**71**, 021706(2005).

CHAPTER 2

Banana-Shaped Liquid Crystals – A Literature Survey

In this chapter, we reviewed the physical phenomena observed in attractive banana shaped mesogenic materials. A survey about banana-shaped liquid crystals of different molecular structures reported in the literature is discussed. This chapter also gives an idea about the current research work in the field of banana shaped liquid crystals.

2. Banana shaped Liquid crystals: it's past and present

The common convention of rod like molecules exhibiting liquid crystalline behaviour has been modified with the synthesis of bent shaped molecules of Schiff base derivatives of resorcinol, by Matsunaga *et al.* [1] and reported only conventional liquid crystalline phases in these bent shaped molecules. However in 1996 the discovery of ferroelectric characteristics in achiral bent shaped molecules presented in **Figure 2.1,** possessing C_{2V} symmetry, which not only exhibit liquid crystalline polymorphism but also distinct switching characteristics, has motivated many of the liquid crystalline scientists to design and synthesize new materials for possible potential applications in emerging opto-electronic technologies. Hence the uncommon distinct molecular structure, the unconventional mesomorphic properties, the occurrence of a spontaneous symmetry breaking in some cases, the biaxial smectic phases and the special switching behaviour of some of their mesophases immediately attracted great interest. Hence, such materials are of great importance because of their possible application in switchable nonlinear optical devices. Bent or banana or V-shaped molecules as shown in **Figure2.1** exhibiting banana mesomorphism has been attracted much attention in the last decade because of its distinct properties [2-10]. The bent shape of the molecules can preferably be packed in a distinct direction viz., the bent direction to manifest the long range correlation of the lateral dipole moments to yield the macroscopic polarization in smectic layers.

2.1. Chirality in Achiral Molecules: ferroelectricity and antiferroelectricity:

Is molecular chirality essential for a liquid crystal to exhibit (anti) ferro electric property?

Takezoe at the 6th International Liquid Crystal Conference 1996 in Kent, Ohio reported, one of the most exciting of the new FLC research areas that burst into prominence, a new type of mesophase called banana liquid crystal phase i.e., achiral banana (or bow or bent) shaped molecules (figure 2.1) exhibiting liquid crystalline behaviour (anisotropic fluid) as well as ferroelectricity (or antiferroelectricity) in liquid crystalline phase with large spontaneous polarization. In 1996 Takezoe, Watanabe and coworkers [2-7] in Japan carried out detailed experimental studies in bent shaped molecules (figure 2.1) to discover that the molecular chirality is not a necessary condition for the existence of ferro or antiferroelectricity in liquid crystalline smectic phases. They discovered that bent or banana shaped molecules exhibit banana liquid crystalline phases and possess ferroelectric order. Moreover this banana shaped achiral molecules [8-10, 12-15] assembling in tilted smectic layers, which manifest the polar order parallel to layers, exhibit distinct ferroelectric properties in the high temperature liquid crystalline phase. This discovery was the beginning of a lot of activity to investigate the structure and properties of these new mesophases, which are not comparable with smectic phases formed by calamitic mesogens. The term "banana-shaped mesogens" is used as a synonym for this new field of liquid crystals; and hence this field has been attracted much attention in recent years because of academic importance and commercial applications.

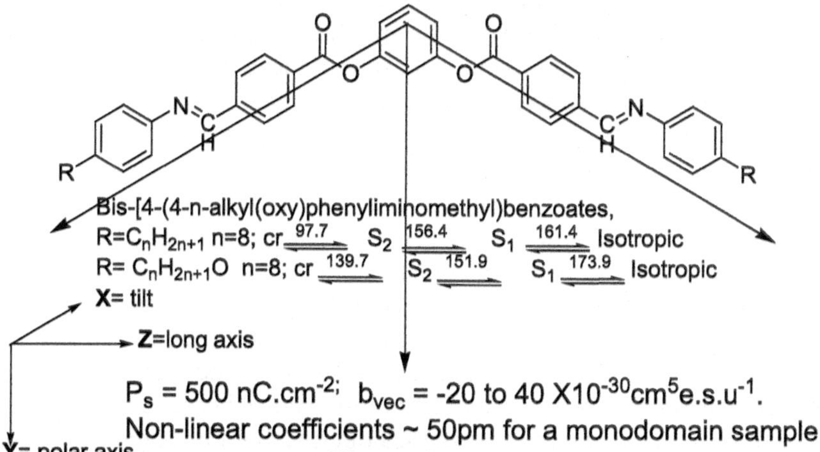

Figure 2.1

So far eight banana phases, B1to B8 have been identified and in two of these phases, B_2 ($S_1$of figure 2.1) and B_5, the polarization direction is inverted in alternating layers, and hence both the phases are antiferroelectric in the absence of applied electric field. The B2 phase is the most common and widely studied phase in bent core liquid crystals. Four different supramolecular packing arrangements are found to exist with two of them can form chiral conglomerates (SmC_AP_A, SmC_SP_F), while the other two appear as racemates (SmC_SP_A, SmC_AP_F) [16, 17]. Ferroelectric switching is also observed in these phases, with a field-induced transition from antiferroelectric to ferroelectric. The observed saturation values for the spontaneous polarization are high (*Ps* =350 nC/cm^2) (and in some compounds as large as 1000 nC/cm^2). In addition, the nonlinear second-order susceptibility for second harmonic generation (SHG) is very large (d_{11} =10 pm/V) and the SHG properties are comparable to inorganic materials. Moreover these materials exhibit liquid crystalline behaviour.

2.2. Possible materials for ferroelectricity:

The necessary requirement for the formation of ferroelectric liquid crystal is the introduction of polar symmetry [11] in a system. Hence the polar symmetry has been obtained by the introduction of the essential requirements viz., chirality of the molecules and their tilted association in smectic layers, leading to reduced symmetry, to form conventional ferroelectric liquid crystals in 1976. However in 1996, the necessary requirements has been replaced by the shape factor, i.e., banana shaped achiral molecules, to induce polar C_{2V} symmetry which has been implicated by the efficient, tilted packing of the banana shaped molecules into a smectic layer, with the direction of spontaneous polarization parallel to two-fold symmetry axis. Thus the tilt of the molecules on the one hand and the polar order of the bent molecules on the other hand are two independent symmetry breaking factors, which cause a chirality of the smectic layers, where the handedness depends on the tilt direction of the molecules. These materials have become potentially promising candidates for basic as well as application oriented research in this decade and shall continue to do so in future as evidenced by the stream of reports appearing in literature currently.

2.3. Design of the molecules:

The possibility of designing and using achiral bent shaped liquid crystalline molecular materials suitable for technical applications in future depends on the understanding of structure–property relationships. Even though a good number of new achiral liquid crystalline molecules are synthesized, few of them have been examined for their electro-optic and NLO characteristics [4-10] a large pool of substances is a very important condition for extensive physical measurements. Few of the important challenges are chemical stability to atmospheric conditions, position of substituents which can affect leading to loss of banana liquid crystalline behaviour, influence of substituents to enhance NLO characteristics etc. The bent shape of these molecules results from a central 1,3-phenylene (or modifying the ring **C** by 2,7-naphthalene [4] unit) and compounds with five, six or seven aromatic (or heterocyclic) rings have been synthesized [12, 13]. Nonetheless, most banana-shaped liquid crystals reported up to now correspond to five-ring mesogens with a 1,3-phenylene central unit as shown in **Figure 2.2**.

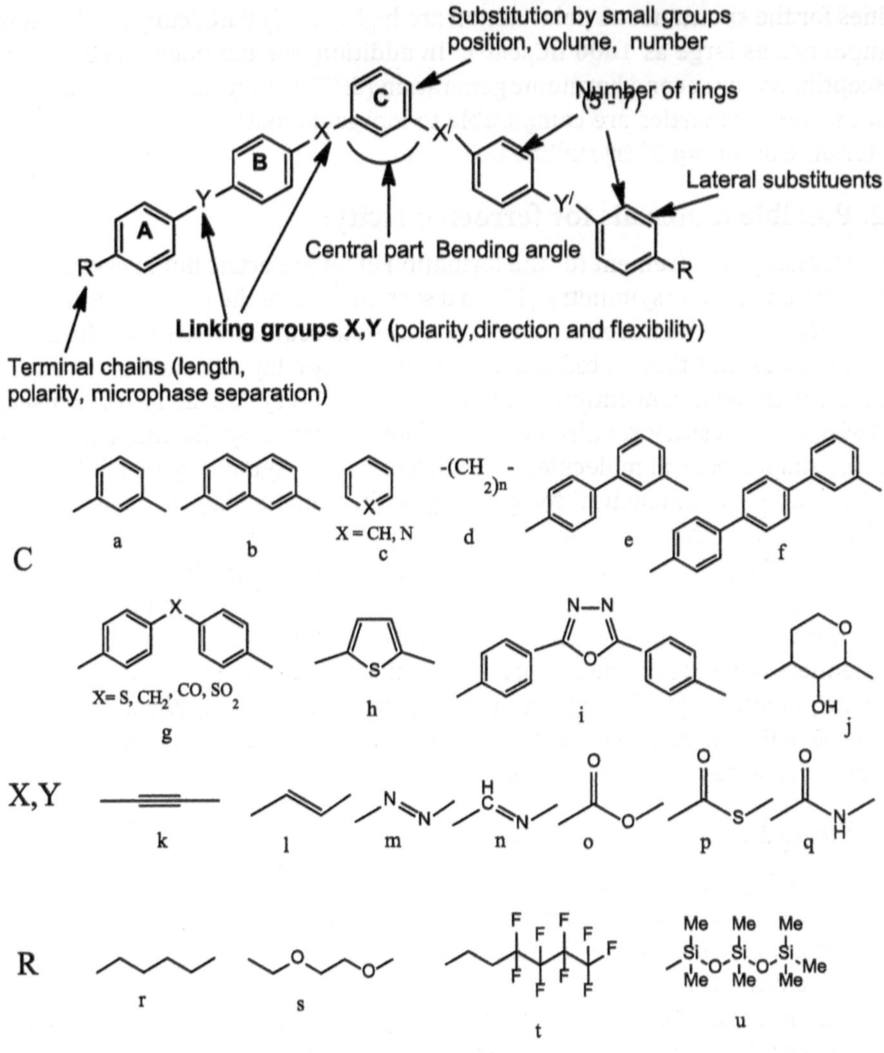

Figure 2.2 : Skeleton of a banana shaped liquid crystal (top). Central bent core C (a-j), linking groups X,Y (k-q) and end chain R (r-u) (bottom) used so far in designing banana shaped liquid crystal.

On the basis of this general formula, different chemical modifications can be considered and according to the available data [12, 13] each alteration can lead to a change in or even to the loss of the liquid crystalline properties. The reasons for this are still not clear. The influence of the nature, position and direction of the linking groups X, Y, X/ and Y/appears very strong. In fact the compounds with azomethine and carboxylic linkages (Y = - N = CH-, X = -COO-, X = -OOC-, Y' = -CH=N) exhibit banana liquid crystalline phases, while the simple inversion of the sense of linking groups, X (-CH=N-) and X'(-N=CH-), leads to the disappearance of these mesophases. It appears that connecting groups corresponding to an alternating distribution of the electronic density throughout the molecules which favour the liquid crystalline properties.

The liquid crystalline properties of the banana-shaped compounds are also strongly influenced by the nature and position of the lateral substituents. However the major drawback in all these compounds exhibiting banana phases is their limited thermal, hydrolytic and photochemical stability because of the presence of the CH=N group. This problem can be overcome either by the introduction of O-hydroxy group with imine linkage promoting to form intermolecular H-bonding to yield a stable molecule or replacing the imine linkage by ester linkage (Y=COO). The physical studies reported in SmC* materials, are being complicated by the hydrolytic instability of the molecules. As pointed out earlier the resorcylidene aniline core actually seems superior to the benzylidene aniline core with respect to exhibiting tilted phases, mesogenicity, and is more stable towards hydrolysis. Recently new achiral electro-optically active bow-phase materials [14, 15] (figure 2.3) are reported, exhibiting potentially useful high strength analog electro-optics in standard LC cells (V-shaped switching). If the linkages *'Y' = resorcylidene and 'X' = -COO-* between the phenyl rings A-B and B-C of figure 2.2 respectively are swapped, the molecule in figure 2.3 becomes a molecule shown in figure 2.4. The O-hydroxyl groups in the ring B of figure 2.4 not only promotes H-bonding and thereby stabilize the compound but can also promote metal complexation to replace the non-liquid crystalline inorganic materials.

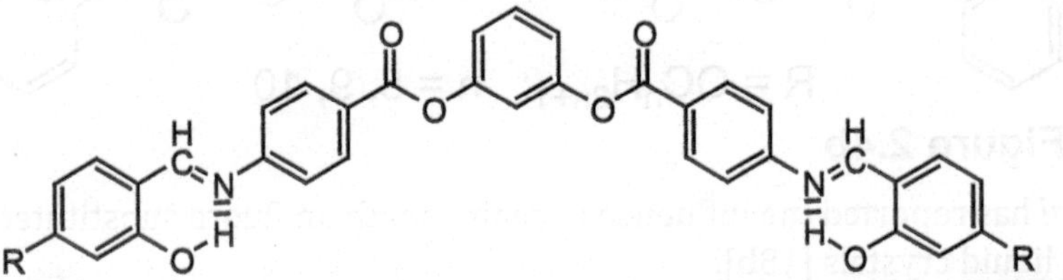

1,3-phenylene bis[N--(2-hydroxy-4-n-alkyloxybenzylidene)-4'amino-benzoates]

R = $C_{10}H_{21}O$ - K $\rightleftharpoons$ (119.4°C) B_2 $\rightleftharpoons$ (180.7°C) I — J.Mater Chem 11, 1818, 2001

R = $C_9H_{19}O$- SmB 2 $\rightleftharpoons$ (120°C / 74°C) SmB 1 $\rightleftharpoons$ (150°C / 145°C) I — J.Mater Chem 11, 2743, 2001

Figure 2.3

Thus, modifying the initial structure with an O-hydroxyl group, attached at position 2 of the phenyl ring B (figure 2.4a) yielded banana liquid crystalline phases. Further the introduction of lateral substituents in the central ring *'C'* like nitro group with its large dipole moment and pronounced electron-attracting ability offers a way to drastically change the electron distribution in the aromatic core. This has been used to enhance NLO-effects such as second harmonic generation (SHG) in liquid crystals and in liquid crystalline polymers. Further, the introduction of a nitro group on position 4 of the central ring (figure 2.4b) produces new mesophases. We synthesized few homologues in our laboratory and the preliminary studies [18] are encouraging. Hence systematic studies are to be performed in a particular homologous series to arrive at the importance of the lateral substituents attached to the central ring. Rao *et al.* have reported the influence of fluoro substituent on 4-position of the central diamine core.

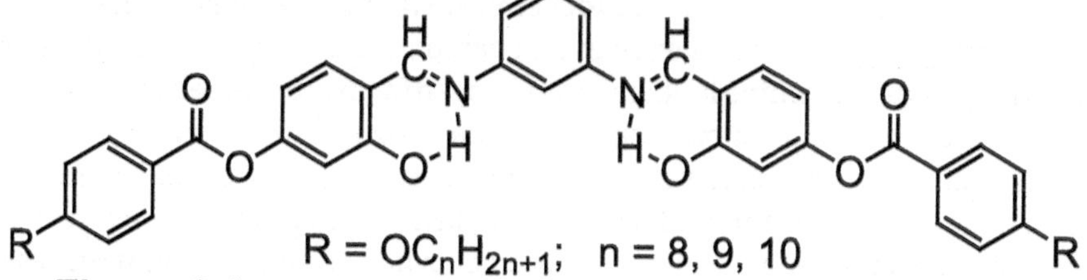

Figure 2.4a

The studies on the influence of substituents on the intermediate or outer rings are very few and meagre [19]. The influence of chain length played an important role in the nature of banana phases exhibited by bent shaped compounds.

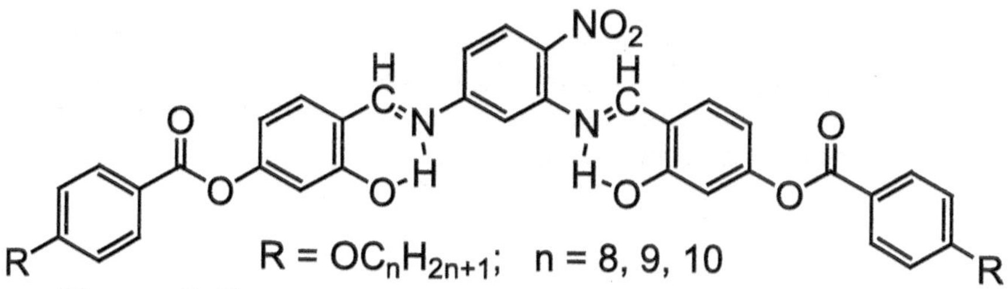

Figure 2.4b

Rao *et al* has reported the influence of chain length on fluoro substituted banana shaped liquid crystals [18b].

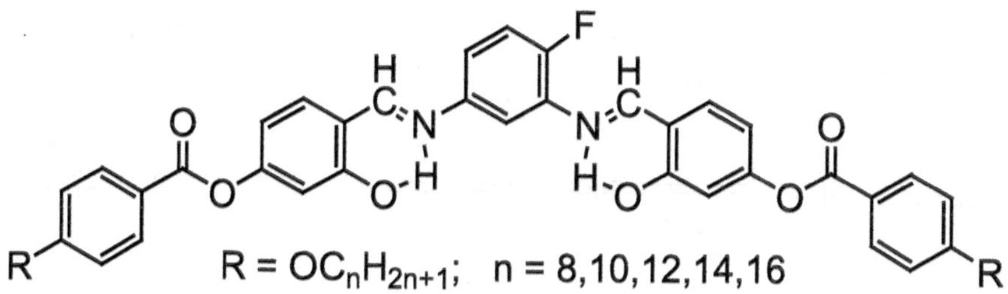

Figure 2.4c

Two or three different mesophases are observed in banana-shaped compounds depending on the length of the alkyl or alkyloxy or alkylthio terminal chains, which is rather subtle and complex unlike that in calamitic mesogens. Nevertheless, in a homologous series, the short homologues exhibit an intercalated banana "B6" and/or a two dimensional "B1 phase" and the long chain derivatives show a switchable lamellar "B2 phase". Other systems present [20, 21] a more complex evolution of the polymorphism depending on the length or on the nature of the terminal chains.

The influence of a lateral fluorine or chlorine or methyl substituents in the terminal or outer phenyl ring gave evidence of the subtle relationship between the chemical structure and mesomorphic properties of banana compounds. Further, lateral substituents can cause steric effects and thereby the Van der Waals volume of the substituents is usually more important with respect to phase behaviour than dipolar interactions. The more detailed influences of lateral substituents depend on the structure of the rigid core and the position of the substituents. Further lateral substitution in calamitic or conventional liquid crystals enhances the molecular width and leads to a larger interpenetration of the terminal chains and decrease in molecular layer spacing. Sometimes the nature of phase also changes. Generalizations are difficult; since the effects differ between polar and non- polar mesogens. Moreover the molecular organization of bent-shaped compounds is a result of a complex and subtle balance between the electrostatic interactions and the van der Waals interactions developed by the aliphatic chains. Thus the distribution of the electrostatic potential maximum (extrema) along the molecule will strongly influence the molecular arrangement and the nature of the mesophases. Hence abundant data is necessary to arrive at generalized conclusions between structure-property relationship and their suitability in practical applications.

In Germany Weissflog*et al*[13] reported antiferroelectric current response while Heppke*et al*[20] observed ferroelectric current dynamics in several other homologues. The discovery of the spontaneous formation of macroscopic chiral domains in smectic phases composed of achiral bow-shaped molecules by Clark, Walba and co-workers [14, 16, 17] in USA has dramatically broadened the scope of the research on chiral/polar smectics. These materials exhibit several completely new phases of matter, and are (based on recent history) quite likely to have more surprises in store. Using detailed electro-optic investigations they gave evidence [14, 16, 17] that polar packed molecules are tilted with respect to layer normal. Most of the materials, which have been reported to exhibit banana liquid crystalline phases invariably, possess Schiff's base imine linkage and a major drawback of these compounds is their limited thermal, hydrolytic and photochemical stability. Hence all the physical studies can be complicated by the hydrolytic instability of the molecules. Even though banana liquid crystalline mesogens, exhibiting ferroelectric and antiferroelectric properties, without Schiff's base units have been reported by Tschierske*et al*, [23-25] variations in the substituents on the bis-Schiff's base bent-core system have proven remarkably useful for providing interesting polar smectics. The resorcylidene aniline core, present in calamitic ferroelectric liquid crystals, actually seems superior to the benzylidene aniline core with respect to mesogenicity, and is more stable towards hydrolysis. This core also played a key role in the first discovery of antiferroelectric phase from achiral molecules, being the core present in the pioneering Soto-Bustamante-Blinov anticlinic bilayer antiferroelectric phase ($SmAP_A$) [11]. Hence introduction of *o*-hydroxyl group to Schiff's base imine linkage not only stabilizes the material to hydrolysis but also enhances the tilted phases due to transverse dipole moment.

2.4. Influence of terminal chain lengths:

The investigation of the relationship between the molecular structure and mesomorphic properties is very important and essential to liquid crystal chemistry. Hence designing a bent shaped molecule with systematic structural variation and synthesis of such designed compounds has to be carried out for systematic investigations in a large number of compounds of different homologous series exhibiting liquid crystalline behaviour to examine the relationship between the chemical structures and their physical properties, particularly the polar orders in the bent-core mesogens. For that purpose, new synthetic methods have to be developed; a large number of compounds are required for any generalization has to be arrived. Excellent reviews [26, 27] appeared recently describing the chiral and polar properties of the compounds, ferroelectric and/or antiferroelectric properties, chirality and optical properties, origin of chirality, molecular structure in different phases, influence of substituents, number of phenyl rings, nature of central core, influence of nature and position of substituent in central core as well as outer rings in the arms of bent molecule.

Since the discovery of polar switching in a bent-core liquid crystal [1] extensive research activity has led to a new era of liquid crystal science particularly in view of polarity and chirality [26, 27]. Lee *et al.* reported [28-32] the emergence of ferroelectricity and antiferroelectricity which is dependent on the number of carbons and position of the chiral along the terminal chains in banana shaped molecules of two homologous series of chiral bent core mesogens viz., 1,3-phenylene bis[4-(4-alkoxyphenyliminomethyl)]benzoates**Figure 2.5a** and their oxygen analogues possessing ω-(S)-amyloxy terminal groups, **Figure 2.5b.**

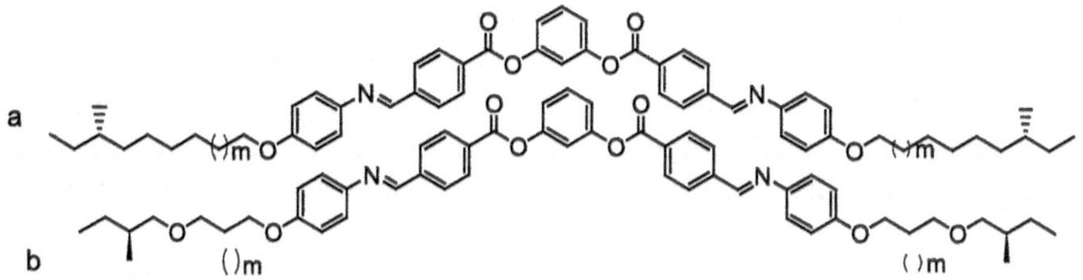

Figure 2.5: 1,3-phenylene bis[4-(4-8-alkoxyphenyliminomethyl)- benzoates]

These achiral molecules spontaneously resolve into chiral domains and exhibit ferroelectric and antiferroelectric phases depending on odd and even of the end chain length in two homologous series (eight compounds) of bent-core liquid crystals. They further demonstrated that the interlayer steric interaction is very important for the selection of polar orders. Since many bent-core mesogens exhibit the antiferroelectric phase, such studies are expected to provide guidelines for the molecular designs to realize the ferroelectric phase.

Tschierskeet *al.* reported [33-36] several bent core compounds **Figure 2.6,** without imine linkage in recent years. They also reported several polyphilic molecules composed of a bent aromatic core, oligo(siloxane) units, and alkyl segments were synthesized, and the self-organization of these molecules was investigated. Most materials organize into polar smectic liquid crystalline phases. The switching process of these mesophases changes from antiferroelectric for the nonsilylated compounds via super paraelectric to surface-stabilized ferroelectric with increasing segregation of the silylated segments.

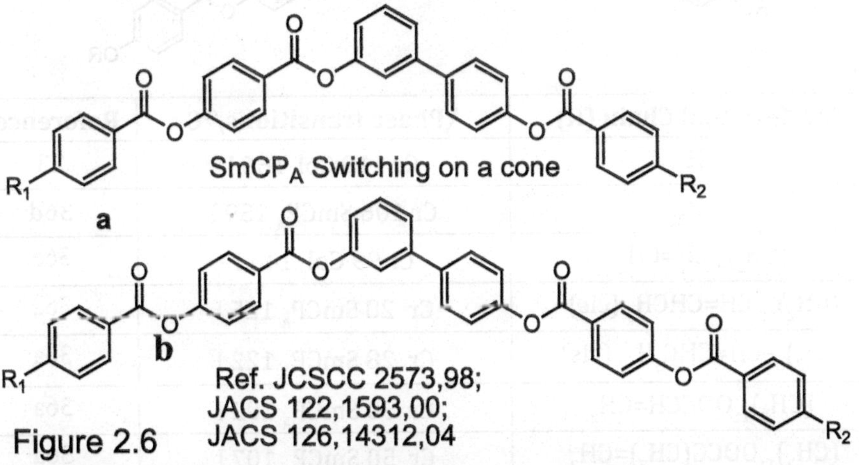

Figure 2.6

Based on the response to electric field they had proposed that the siloxane sublayers stabilize a polar synclinic ferroelectric (SmCsP$_F$) structure, and the escape from a macroscopic polar order as well as steric effects leads to a deformation of the layers with formation of disordered micro-domains, giving rise to optical isotropy. Another striking feature is the spontaneous formation of chiral domains with opposite handedness. For two compounds, a temperature-dependent inversion of the optical rotation of these domains was found, which is associated with an increase of the tilt angle of the molecules from <45° to >50°. This observation confirms that it is a new source of optical activity in supramolecular systems. Further they also observed phase segregation is lost with the increasing alkyl chain length and a transition from smectic to a columnar phase. In the columnar phase, the switching process is antiferroelectric and takes place by rotation of the molecules around the long axes, which reverses the layer chirality; that is, the racemic ground-state structure is switched into a homogeneous chiral structure upon application of an electric field.

The presence of cis double bonds (Table 2.1) which increase the flexibility of the chains, does not decrease the clearing temperatures significantly, but drastically reduces the melting points, leading to bent-core mesogens with glassy polar smectic phases[36]. Also branching of the chains is of great importance and can give rise to low melting points and glassy materials [36b]. Even more importantly, branching can lead to a transition from AF to FE smectic phases [4b] as well as to B7 subtypes due to changes at the interlayer interfaces.

Table 2.1: Comparison of the mesophases shown by compounds containing different types of terminal chains

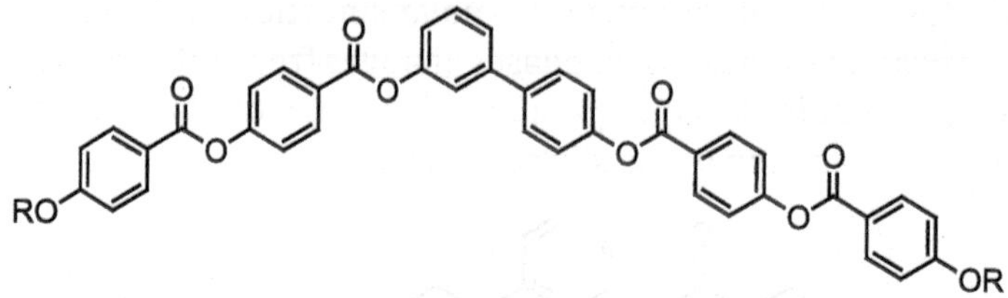

The Terminal Chain (R)	\Phase transitions/°C	References
$C_{10}H_{21}$	Cr 119 Col_r 166 I	23
$C_{12}H_{25}$	Cr 106 $SmCP_A$ 159 I	36d
$(CH_2)_9$ $CH{=}CH_2$	Cr 99 Col_r 146 I	36e
$(CH_2)_{10}$ $CH{=}CHCH_3$ (cis)	Cr 28 $SmCP_A$ 125 I	36a
$(CH_2)_5{-}CH{=}CHC_5H_{17}$ (cis)	Cr 28 $SmCP_A$ 122 I	36a
$(CH_2)_{11}OOCCH{=}CH_2$	Cr 50 $SmCP_A$ 102 I	36a
$(CH_2)_{11}OOCC(CH_3){=}CH_2$	Cr 50 $SmCP_A$ 107 I	36a
$(CH_2CH_2O)_4CH_3$	Cr 22 $SmCP_A$ 62 I	36a
$(CH_2)_{11}SiMe_2OSiMe_2OSiMe_3$	Cr 87 $USmCP_A$142 $Col_{ob}P_A$155I	36c
$(CH_2)_6C_4F_9$	Cr 127 $SmCP_A$ 213 I	23

2.5. Influence of linking groups in bent core molecules:

Weissfloget *al.* [37] has reported the role of direction of the linkage groups on banana shaped liquid crystals as detailed in **Table 2.2**. For this reason they have synthesized two series of isomeric five-ring bent core mesogens with equal length of the terminal chains (octyloxy and dodecyloxy), which contain only carboxyl linkage groups. All possible 10 isomers are presented which are different from each other by the direction of at least one linkage group. The mesophase behaviour and the physical properties vary significantly and the influence of direction of the carboxyl linkage groups in both series of isomeric compounds is demonstrated. The direction of connecting groups between aromatic rings in calamitic liquid crystals can also influence the mesophase behaviour, but not as prominent as in these compounds.

Table 2.2: Influence of direction of ester linkage on mesophase behaviour.

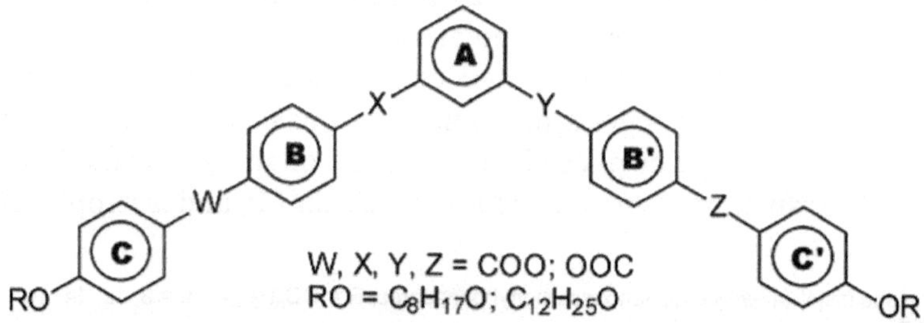

W	X	Y	Z	Cr		SmCP		Col		I
COO	COO	OOC	OOC	●	109	●	119	-		●
COO	OOC	COO	OOC	●	157	-		-		●
OOC	COO	OOC	COO	●	152	-		●	162	●
OOC	OOC	COO	COO	●	191	-		●	189	●
COO	COO	COO	COO	●	105	-		●	140	●
COO	COO	COO	OOC	●	120	●	112	-		●
COO	COO	OOC	COO	●	110	●	133	-	168	●
COO	OOC	COO	COO	●	142	●	158	●	144	●
COO	OOC	OOC	COO	●	146	-		●	168	●
OOC	COO	COO	COO	●	138	-		●		●

Weissfloget *al.* [37] also reported novel five ring bent core systems possessing only ester connecting groups between the aromatic phenyl rings and different lateral substituents at the central phenyl ring viz., 1,3-phenylene bis[4/-(4//-alkyloxyphenoxycarbonyl)benzoates**Figure 2.7a** and isomeric 1,3-phenylene bis-[4/-(4//-dodecyloxybenzoyloxy)benzoates **Figure 2.7b** with alkyl chain length of octyloxy or larger are presented. They demonstrated that the influence of lateral substituents on the banana mesophase behaviour is much more pronounced than in the calamitic compounds.

They also found that mesophase behaviour strongly depends on the position of the lateral substituents. Compounds, which are derived from 4-cyano-, 4-chloro- and 4,6-dichloro-resorcinol, show polymorphism variants where polar phases (SmAP, SmCP) occur together with nematic and conventional smectic phases, e.g. SmA–SmAP, SmA–SmC_SP_A–Col_{ob}–SmC_SP_A, N–SmA–$SmCP_A$, SmA–SmC–$SmCP_A$ and SmC–$SmCP_A$. On the basis of the behaviour of two series of materials, the occurrence of different polar-switching mechanisms had been demonstrated. Apart from the usual mechanism by director rotation around the tilt cone, the polar switching can also take place through collective rotation of the molecules around their long axes in $SmCP_A$ and $SmCP_F$phases, which corresponds to a field-induced switching of the layer chirality viz., coupling of reversal of layer chirality with field reversal. A remarkable finding is the polar switching in the crystalline modification of long-chain, bent-core compounds with a methyl group in 2-position, which is accompanied by a clear change of the optical texture and by a relatively high switching polarization (approx.600 nC/cm^{-2}). It was found for selected laterally substituted bent-core compounds that, above the transition temperature of a polar to a non-polar phase, the non-polar phase can be transformed to the polar phase by application of an electric field, which was proved for the transitions isotropic–SmCP$_F$, SmA–SmCP$_F$ and isotropic–CrII polar phase.

1,3-phenylene bis[4′-(4′′-alkyloxyphenoxycarbonyl)benzoates, R = OC_nH_{2n+1}; n = 8, 12, 14, 16

Figure 2.7a

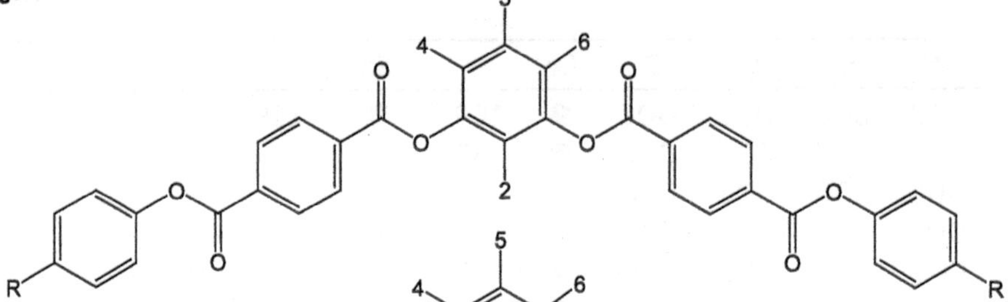

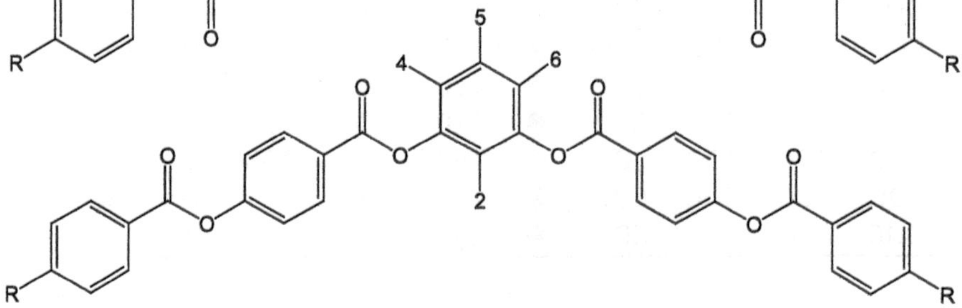

1,3-phenylene bis-4′-[4′′-dodecyloxybenzoyloxy)benzoates R = OC_nH_{2n+1}; n = 12

Figure 2.7b

Recently Gomola*et al.* [38, 39] reported two different types of homologous series of compounds **Figure 2.8 and 2.9,** exhibiting orthogonal polar smectic A phases. Three series of asymmetric bent-core compounds were synthesized and their mesomorphic properties were studied. All the compounds possess 2-methylresorcinol as a central unit, nitro and/or chloro substituents in the outermost rings and a stilbene linking group between in two outer phenyl rings as a main element in the structure. Molecules of series 1 and series 2 possess only saturated terminal chains while those of series 3 have one or both unsaturated chain/s. All the compounds exhibit the isotropic to uniaxial orthogonal phase with ferroelectric properties SmAP$_R$ phase transition and the B1$_{RevTilted}$ phase at lower temperatures. For compounds with one olefinic terminal chain, the temperature range of the SmAP$_R$ phase was found to be narrower in comparison to compounds with both saturated terminal chains. Dielectric investigation inferred that both the SmAP$_R$ and B1$_{RevTilted}$ phases and two different types of dynamical behaviour were observed.

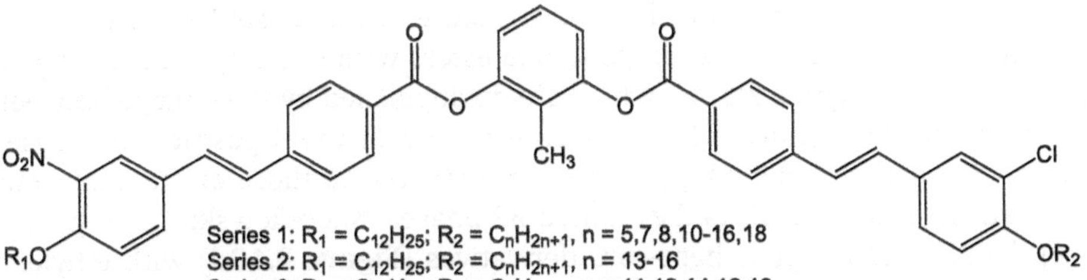

Series 1: $R_1 = C_{12}H_{25}$; $R_2 = C_nH_{2n+1}$, n = 5,7,8,10-16,18
Series 2: $R_1 = C_{12}H_{25}$; $R_2 = C_nH_{2n+1}$, n = 13-16
Series 3: $R_1 = C_{11}H_{23}$; $R_2 = C_nH_{2n+1}$, n = 11,12,14,16,18

Figure 2.8

The symmetric compounds with 2-acetyl groups exhibited polarization randomised non-tilted smectic A ($SmAP_R$) and room temperature non-tilted smectic ($SmAP_A$) phases, a first example of polar biaxial Smectic A to polar uniaxial smectic A phase.

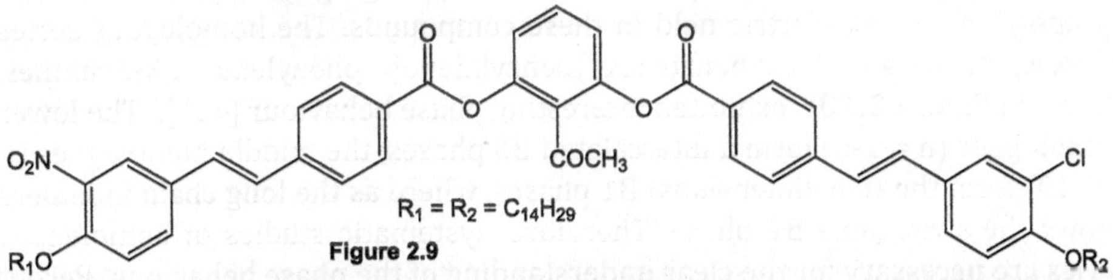

$R_1 = R_2 = C_{14}H_{29}$

Figure 2.9

Nguyen *et ul.* [40], Weissflog*et al.* [41] reported the five ring banana shaped molecules derived from isophthalic acids, as shown in **Figure 2.10b** and **c**, apart from different substituted compounds (**Figure 2.10**a). By a combination of several linking groups and different selected substituents either on the outer ring or on the central ring several mesophases with switching properties are detected. Their studies revealed new structural modifications and also the importance of alkyl chain length on banana polymorphism. They demonstrated the importance of the nature of lateral substituent and its position on the central ring in affecting the liquid crystalline properties. Further they also reported the importance of the nature and the sense of the linking groups and also the sense of bridging group of the central linking group in deciding the rigidity of the molecule and conformational freedom in exhibiting switchable banana phases. They reported B1, and B6 phases in lower homologues and B2 phase higher homologues in bent shaped molecules with a thio ester linkage. They also reported ferroelectric switching behaviour for a B7 phase in fluoro substituted bent-core molecules in peripheral positions on the aromatic core, ortho to each of the two terminal alkoxy chains. This special substitution pattern seems to be a quite useful structural unit for the formation of ferroelectric switchable banana phases, which was evidenced in fluoro substituted compounds [42, 43]. Nguyen *et al.* reported [40g] derivatives of esters of isophthalic acids (**Figure 2.10b**), which exhibited conventional Sm C and nematic phase, while the compounds substituted with a fluorine atom (X= F) exhibited switchable phases. The fluorine derivatives exhibited four smectic phases without in-plane order but the structure of the mesophase are not yet known. The derivatives of isophthalic acid esters with five-ring banana-shaped mesogens 41a] (**Figure 2.10c**), which are distinguished by the linkage between the aromatic rings and the Bromine substitution in the 4-position, decreases the melting and clearing temperatures as compared to those of the analogous unsubstituted compounds and exhibited B2 phase. Mieczkowski*et al.* reported [44] the synthesis of new bent-core derivatives (**Figure 2.10d**) with ethylene linkage in banana branches instead of an imine or ester linkage and incorporating lateral substituents like fluorine or iodine atoms. The compounds exhibited B7 and B2 phases with anticlinic (SmC_AP_A) or synclinic (SmC_SP_A) interlayer structures, depending on the central ligand and lateral substituent and also found lower transition temperatures than the unsubstituted analogues. The SmC_AP_A phase exhibited strong optical activity and chiral domain segregation could be induced by applying a weak electric field in these compounds. The homologous series of N,N/-bis[4-(4-n-alkoxybenzoyloxy)benzylidene] phenylene 1,3-diamines, shown in **Figure 2.10e**, exhibited interesting phase behaviour [41b]. The lower homologues (n = 1-6) form a intercalated B6 phases, the middle homologues (n = 5-10) form the two dimensional B1 phases, where as the long chain members shows the switchable B2 phase. Therefore, systematic studies of homologues series are necessary for the clear understanding of the phase behaviour. Pelzl*et al.* reported [41c] a novel chiral banana shaped mesogens, 4-choloro resorcinol bis[4-(4-n-dodecyloxy benzoyloxy) benzoate] (**Figure 2.10f**), exhibiting nematic phase with unusual optical and electro-optical properties. First examples of achiral banana shaped compound (**Figure 2.10g**) possessing perfluoroalkyl chain in the terminal position and lateral substituent at the top (or 5-position) of the central benzene ring exhibited B2 and SmA phases, characteristics for a bent and rod shaped molecules with reduced clearing and melting temperatures have been reported by Kovalenko *et al.* [40]. Pelzl*et al.* reported [41d] the first banana phase, B6, found in Vorlander's achiral banana shaped compounds synthesized during 1900-37, which is shown in **Figure 2.10h.**

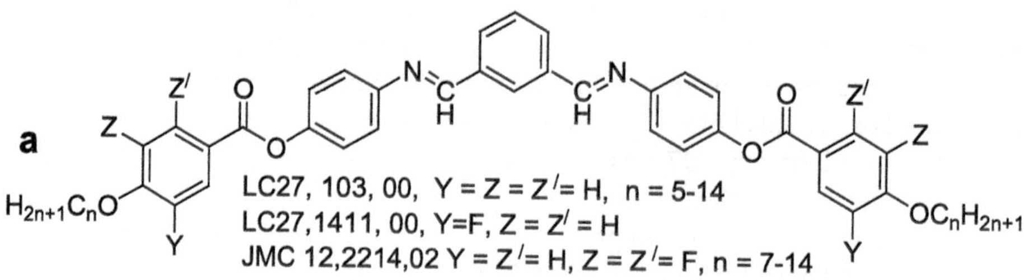

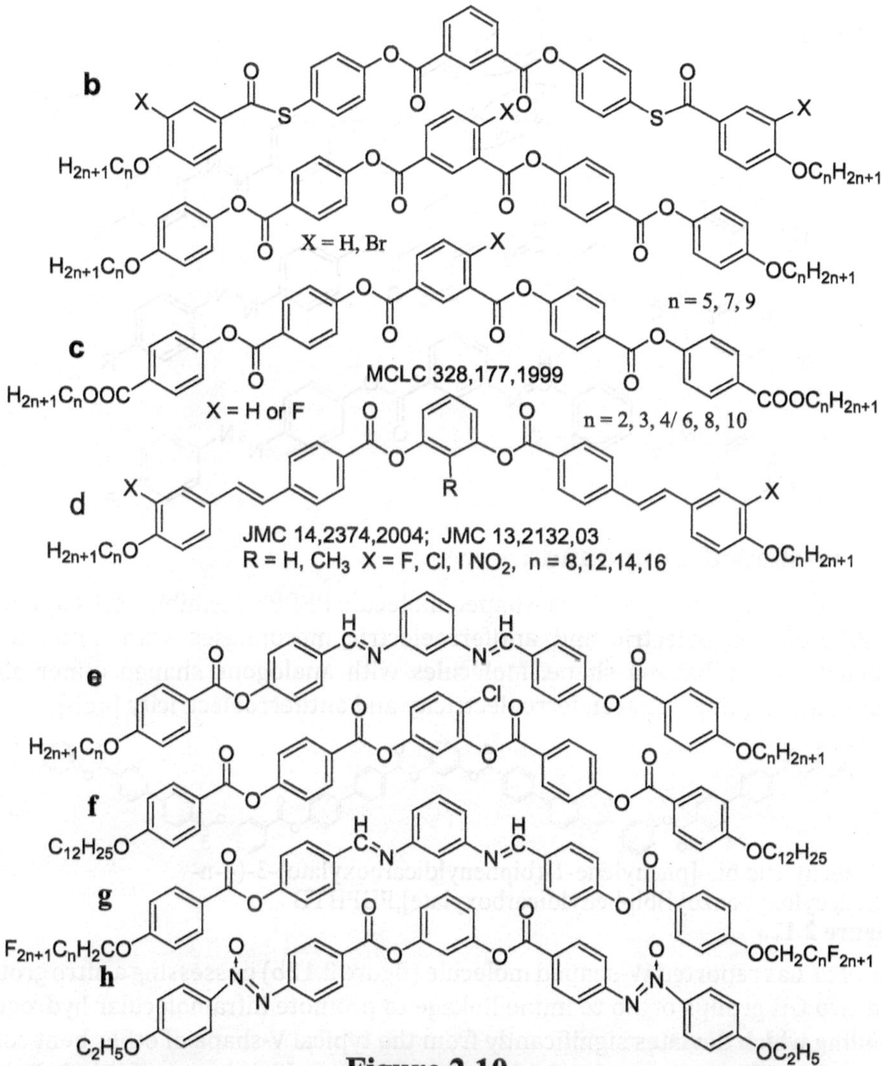

Figure 2.10

Prasad *et al.* reported [45] first examples of achiral bent-core molecules (**Figure 2.11**) of five aromatic rings, linked through an azo linkage and with an imine and/or ester linkage, exhibiting B1 and B2 banana mesophases relatively at low transition temperatures and with wide temperature ranges. The B2 phase of these materials is identified to be the anticlinic–antiferroelectric, SmC_AP_A phase The observation of bent-core phases in azo compounds assumes significance from the fact that the introduction of the –N=N– linkage adds a new dimension, namely photochromicsm, to this field.

Figure 2.11

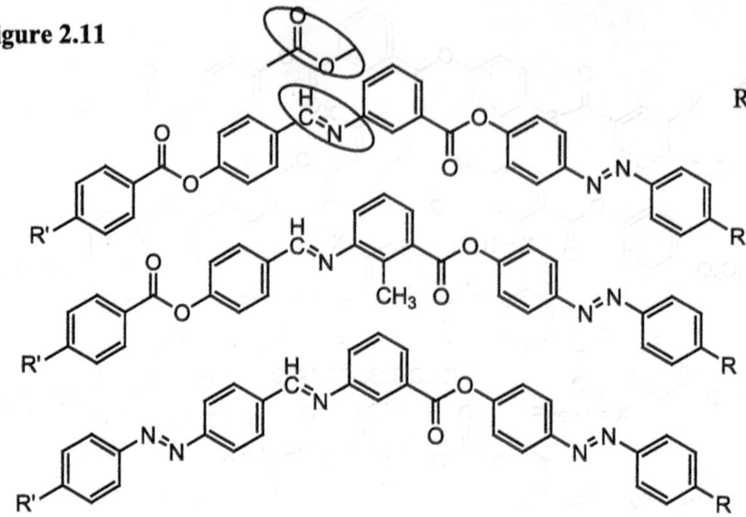

Ref. 45

2.6. W-shaped compounds:

Kumar *et al.* reported [46a] a W-shaped molecule PBPBTB exhibiting a sequence of different ferroelectric and antiferroelectric mesophases with anticlinical coupled double banana shape. Molecules with analogous shaped dimer also exhibit a switching between ferroelectricity and antiferroelectricity [46b].

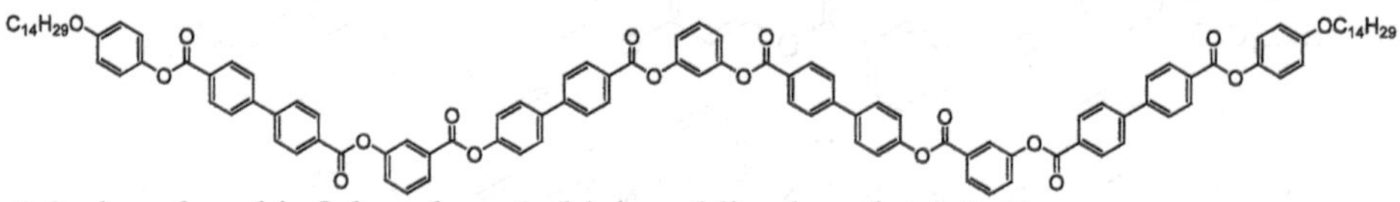

1,3-phenylene bis-[phenylene-1-(biphenyldicarboxylato)-3-(4-n-tetradecyloxybenzoyl)biphenyldicarboxylate],PBPBTB
Figure 2.12a

Rao *et al.* has reported W-shaped molecule (figure 2.12b) possessing a nitro group and two OH groups ortho to imine linkage to promote intramolecular hydrogen bonding,which deviates significantly from the typical V-shape of other bent core molecules. The compound exhibits chiral filament like domains[47] and also chiral domains of opposite handedness which can be seen in this birefringent phase by slightly decrossing the polarizers. This compound exhibits an unusual smectic mesophase with additional in plane order (Sm X2).

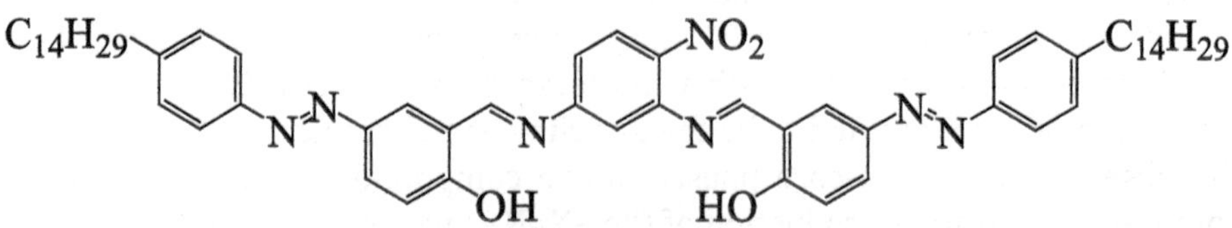

Figure 2.12b

2.7 Calamitic compounds with a spacer:

Watanabe *et al.* reported [48, 49] interesting mesophases, as found out by their banana leaf like textures (($(C_2H_2)_5$ spacer), exhibited by the dimeric systems (**Figure 2.13a**) depending upon the number of carbon atoms (parity) in the spacer joining the two aromatic fragments with alkyl or alkoxy end chains. Further such achiral systems exhibit antiferroelectric or nonferroelectric switching behaviour under the influence of triangular wave electric field provided the odd alkylene spacer (promoting bent shape) and tail parts of the dimer segregate to form smectic layers. Moreover second harmonic generation studies [48] confirmed the growth of macroscopic polar order under the influence of electric field in dimeric molecules. The homologous series of symmetric dimeric liquid crystals, where two two-ring or three-ring mesogenic units are connected by a bis (carbonyloxy) propylene spacer (**Figure 2.13b, c**), exhibited conventional calamitic smectic phases [50]. The most interesting result is the inverse phase sequence SmA to SmC has been observed with increasing temperature in compounds of three ring mesogenic units. Ferroelectric switching behaviour is also reported [51] in bent shaped mesogens (**Figure 2.13d**) possessing non-mesogenic units linked by alkylene spacer by interchanging the ester linkage with imine linkage unit in the nonmesogenic unit.

Figure 2.13

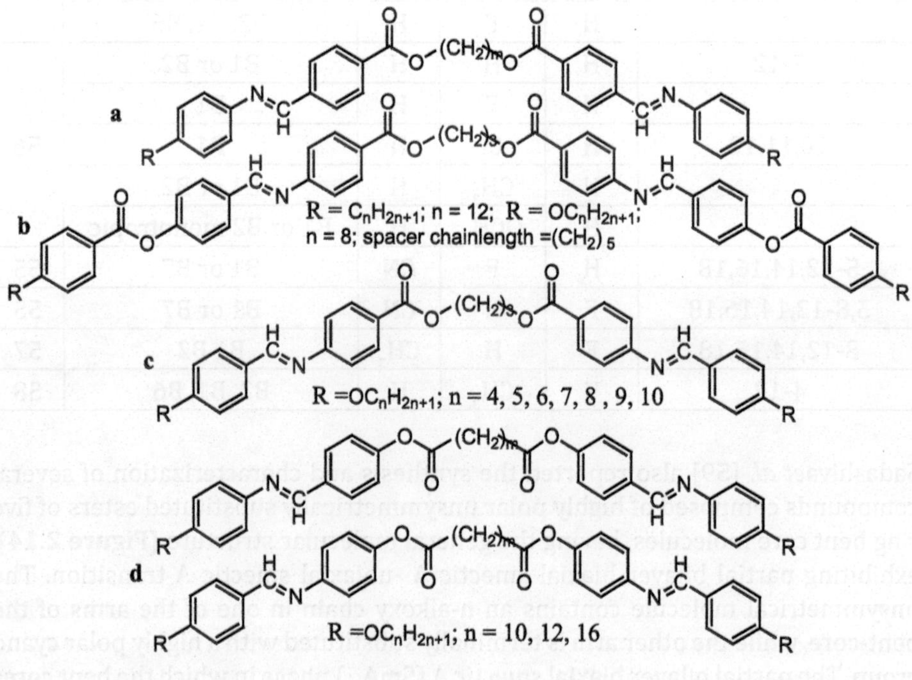

2.8. Effect of substituents on banana shaped mesogens:

Sadashiva and his coworkers reported several compounds exhibiting banana liquid crystalline behaviour during the last few years. Sadashivaet *al.* [52-58] first reported the effect of a number of lateral substituent on the mesophase formed by compounds composed of achiral banana-shaped molecules viz., derivatives of resorcinol possessing seven aryl rings with ester linkages. The synthesized compounds and the reported phases are shown in **Table 2.3**.

Table 2.3

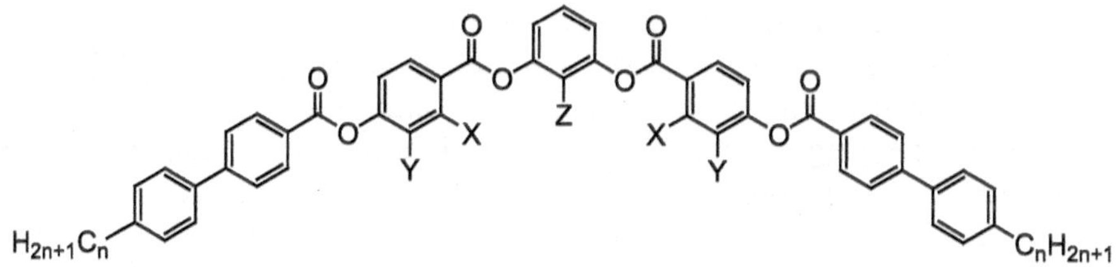

Chain length (n)	X	Y	Z	Phase sequence	Ref.
	H	H	H	B2, B1, B6	
	F	H	H	B2, B1, B6	
5-12, 14,16,18	CH_3	H	H	B2, B1, B6	53
	C_2H_5	H	H	B2, B1, B6	
	H	F	H	B2, B1, B6	
7-12	H	H	H	B1 or B2	
	H	F	H	B1	
10,11,12	H	Cl	H	B1	56
	H	CH_3	H	B1 or B2	
	H	OCH_3	H	B1 or B2 monotropic	
5-12,14,16,18	H	F	CN	B1 or B7	55
5,8-12,14,16,18	F	H	CN	B1 or B7	55
8-12,14,16,18	F	H	CH_3	B1 B2	57
4-12	H	CH_3	H	B2, B1, B6	58

Sadashivaet *al.* [59] also reported the synthesis and characterization of several compounds composed of highly polar unsymmetrically substituted esters of five ring bent core molecules, having the general molecular structure **(Figure 2.14)**, exhibiting partial bilayer biaxial smectic A -uniaxial smectic A transition. The unsymmetrical molecule contains an n-alkoxy chain in one of the arms of the bent-core, while the other arm is terminally substituted with a highly polar cyano group. The partial bilayer biaxial smectic A (SmA_{db}) phase in which the bent cores of two neighbouring molecules overlap in an antiparallel orientation as in the case of highly polar lath-like molecules.

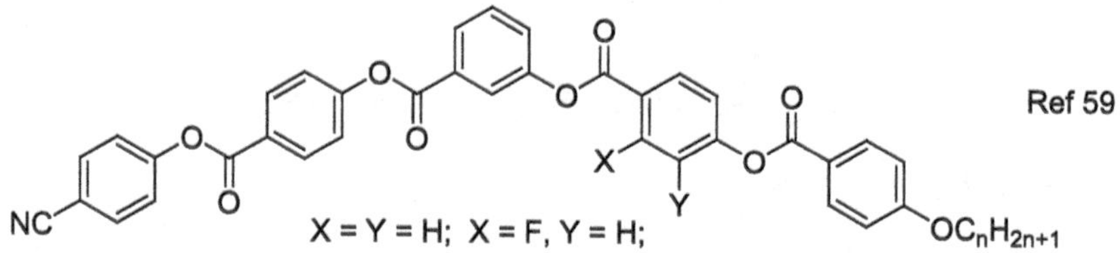

X = Y = H; X = F, Y = H;
X = H, Y = F; n = 12-18

Figure 2.14

Reddy *et al.*, Thisayukata*et al.* reported the homologous series of bent shaped compounds containing 2,7-dihydroxynapthalene [5,53,60] and 1,3 dihydroxybenzene (resorcinol) [60] as a central bridging unit with lateral fluoro substituent(s) on the inner or outer phenyl rings. Some of the homologous compounds exhibit switchable B2 phase. The lower homologous compound display only nematic phase and B1 phase where as higher homologous compound exhibits B2 and monotropic B3 phases or columnar phases depending upon the number and position of fluoro substituents. The lateral fluoro-substituents not only lower the clearing transition temperature of the parent compound but also in lower homologous series displays monotropic nematic phase. The compounds with lateral fluoro substituent meta to ester linkage of the end phenyl ring exhibit B3 phase. The work can be summarized as follows. They have synthesized derivatives of resorcinol and 2,7-naphthol with either two or three rings in the wings of the molecule and studied structure-mesomorphism relationship with different substituents either on the central core ring or in the rings of the wing as shown in **Table 2.4**. The salient features are that some of these compounds are to exhibit ferroelectric properties, chiral behavior, biaxial phases, columnar phases and, calamitic phases depending on the position and nature of substituent. All the compounds are esters and are thermally stable. Several novel columnar (switchable and non switchable), ferroelectric and anti ferroelectric phases have been reported in the banana shaped mesogens derived from symmetric 2,7-dihydroxynaphthalene and 1,3-dihydroxyderivatives (**Table 2.4**) [53, 60].

Table 2.4

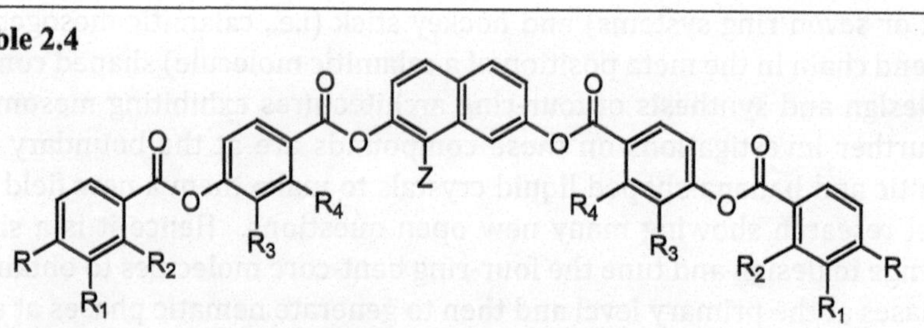

Chain length(R)	R1	R2	R3	R4	Z	Phase variants	Ref
n = 5-12, 14, 16, 18	H	H	H	H	H	B1 or B2	60
n =1-12, 14, 16, 18	H	H	H	F	H	B1,N or B2,	
n =4- 12, 14,16, 18	H	H	F	H	H	B1,N or B3, B2,	60
n = 7-9	H	H	H	H	H	B1	61

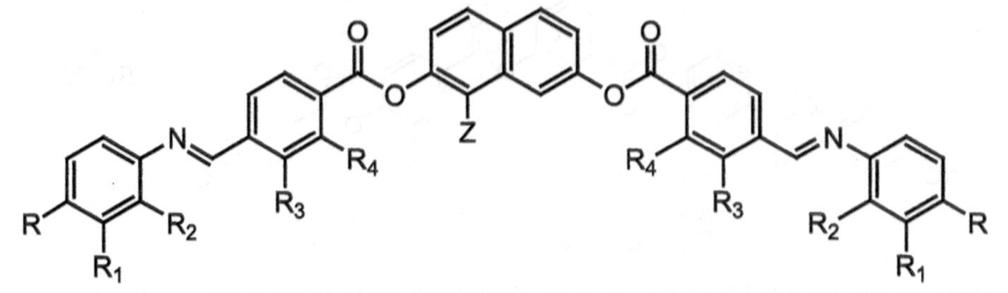

Chain length(R)	R1	R2	R3	R4	Z	Phase variants	Ref
n = 5-12,14,16,18,6-Me-8*	H	H	H	H	H	Sm3, Sm2, Sm1	3, 5
n = 6, 8, 10,12	H	H	H	H	H	BX,	62
n = 6, 8, 10,12	H	H	H	H	CH_3	B2,	62
n = 6, 8, 10,12	H	H	H	H	CN	B2/-B2//-N	62
n = 6, 8, 10,12	H	H	H	H	Cl	B2	62

2.9. Four-ring banana shaped liquid crystals:

The reported results on four-ring bent-core compounds till date are meagre and the general idea to be derived from any specific report in literature to understand or speculate their mesophase behaviour is lacking. The reported results pointed towards the classical phases exhibited by calamitic compounds and hence the speculation suggested that they may not be able to form typical banana-phases (SmAP, SmCP) spontaneously because the required shape/steric as well as core-core interactions between bent molecules are too small and are not sufficient enough to induce banana mesomorphism. Hence the four-ring bent-core molecules has been classified as intermediate between the banana-shaped (five or six or seven ring systems) and hockey stick (i.e., calamitic mesogens with a alkyl end chain in the meta position of a calamitic molecule) shaped compounds. The design and synthesis of four-ring architectures exhibiting mesomorphism and further investigations on these compounds are at the boundary between calamitic and banana-shaped liquid crystals to make them a new field of liquid crystal research showing many new open questions. Hence it is a significant challenge to design and tune the four-ring bent-core molecules to obtain banana LC phases at the primary level and then to generate nematic phases at a distinct level.

Kang *et al.* [63] reported bent core mesogen derived from oxadiazole **(Figure 2.15)**as central core with a bent angle of ~134^0 which is larger than the common bent angle of 120^{0}found in 1,3-substituted phenyl or 2,7-substitueted naphthalene derivatives exhibiting LC polymorphism. This compound exhibits a nematic, a SmC and an optically isotropic uniaxial Bx phase with randomly distributed domains of opposite handedness. The Bx phase under the influence of electric field transformed into a birefringent phase with homochiral SmC_AP_F structure.

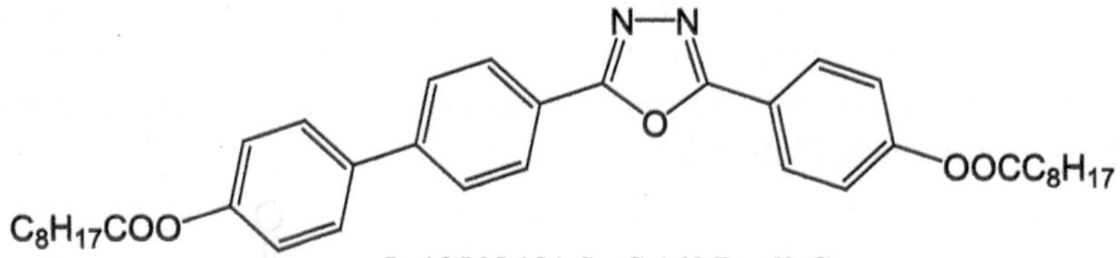

Is 185 N 181 SmC 160 Bx 69 Cr

Figure 2.15

The four-ring compounds that had been reported in literature consisting of four aromatic phenyl rings in a bent core are compiled in Table 2.5. Majority of these four-ring compounds are unable to exhibit liquid crystalline behaviour, while very few of them are mesogens.

Table 2.5: Chemical structure and mesophase behaviour of bent core four ring mesogens reported in the literature

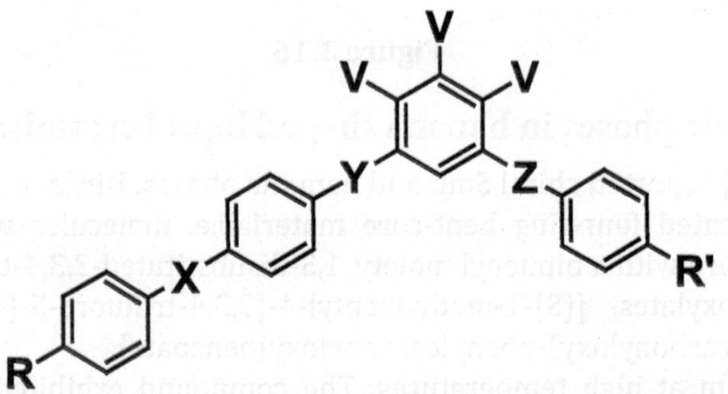

R	X	Y	Z	V	R/	Mesophase behaviour	Ref.
$C_{12}H_{25}O$-	COO	CH=CH	OOC	H	$-OC_{12}H_{25}$	Cr 108 Is	64
NC-	---	OOC	OOC	H	$-OC_{10}H_{21}$	Cr 141 Is	65
$C_7H_{15}O$-	COO	COO	COO	H	$-OC_8H_{17}$	Cr 130 Is	66
$C_{10}H_{21}O$-	CH=CHCOO	CH=N	OOC	H	$-OC_8H_{17}$	Cr 96 Is	
H	-N=N-	N=N	OOC	H	$-OC_{12}H_{25}$	Cr 107 Is	67
$C_{10}H_{21}O$-	--	OOC	OOC	F	$-OC_{12}H_{25}$	Cr 88 Is	68
$C_{12}H_{25}O$-	COO	COO	OOC	F	$-OC_{12}H_{25}$	Cr 70(N41)	69
$C_{12}H_{25}O$-	---	COO	COO	F	$COOC_8{}^*H_{17}$	Liq.Cryst. [a]	68

[a]Cr 99 $SmC_a{}^*$ 137 SmC^*_{ferri} 141 SmC^*_{ferro} 166 SmA* 213.5

Weissflog*et al.* reported novel asymmetric four-ring bent-core compounds derived from N-benzoylpiperazine (figure 2.16) exhibiting SmA, SmC and SmI phases. In compounds exhibiting SmA and SmC phases polar structures can be induced under the influence of external electric field. Some compounds exhibit SmI phases, which are found to be polar ferroelectric phases and are switchable under the influence of electric field. This is the first example of non-chiral ferroelectric SmI phase (Figure 2.16).

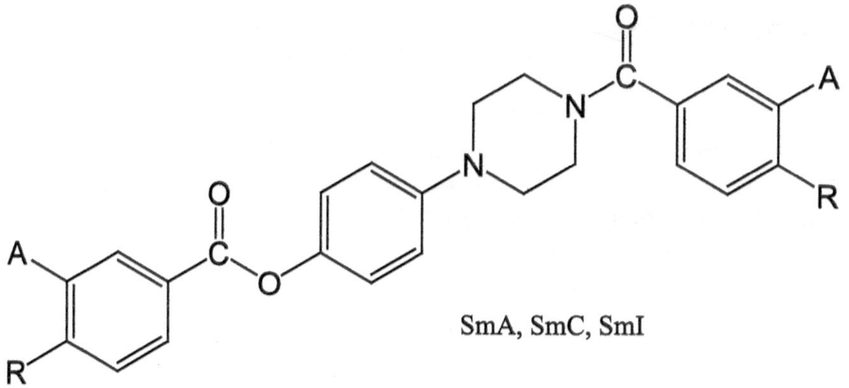

Figure 2.16

2.10. Nematic phases in banana shaped liquid crystals:

Hird*et al.* [70] reported chiral SmC and nematic phases. Hird*et al.* reported [71] fluoro substituted four-ring bent-core materiali.e. molecules with only four rings in the core with a biphenyl moiety, 1,5-disubstituted-2,3,4-trifluorophenyl biphenylcarboxylates [(S)-1-methylheptyl-4-[2,3,4-trifluoro-5-(4/-dodecyloxy-biphenyl-4-ylcarbonyloxy)-phenylcarbonyloxy]benzoate] exhibiting mesomorphism at high temperatures. The compound exhibited conventional SmA, ferro-, ferri-, and antiferroelectric SmC phases resulting from the molecular chirality and not due to the bent shape.The compound with conventional directions of ester groups based on banana-shape of parent compound is non-mesogenic, with a rather high melting point which may be masking liquid crystal phases and further the bend position is in not the right place for conventional mesomorphism. Increase in chain length and modification in the direction of linking ester moiety revealed the monotropic nematic phase at a lower temperature. The nematic phase exhibited Very unusual optical textures and when viewed between crossed polarizers exhibited domains of opposite handedness present in equal proportions. The electro-optic switching of the nematic phase induces a fan-like texture that shows an unusual switching behaviour. Further no homeotropic texture could be observed in the nematic phase. Structures of the banana shaped liquid crystals reported which shows nematic phases are presented below (figure 2.17).

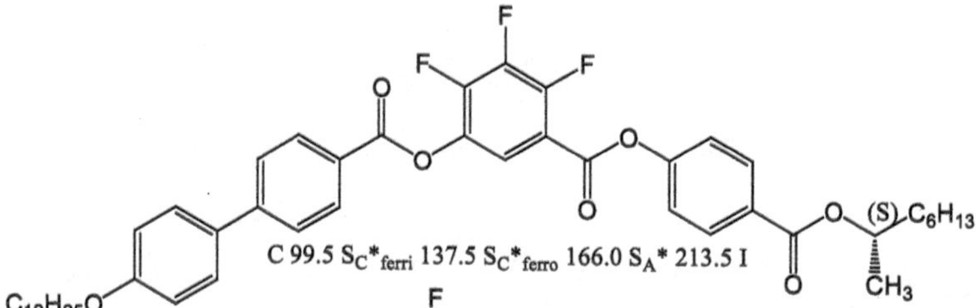

C 99.5 $S_{C}{}^{*}{}_{ferri}$ 137.5 $S_{C}{}^{*}{}_{ferro}$ 166.0 $S_{A}{}^{*}$ 213.5 I

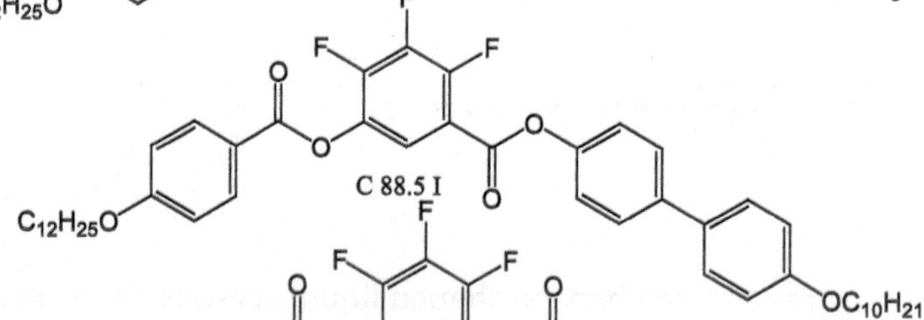

C 88.5 I

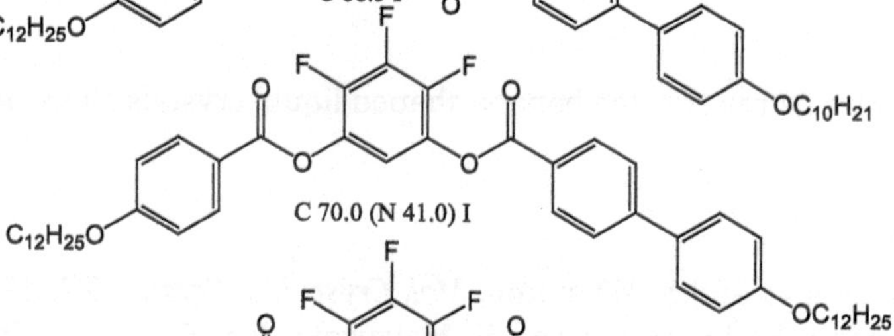

C 70.0 (N 41.0) I

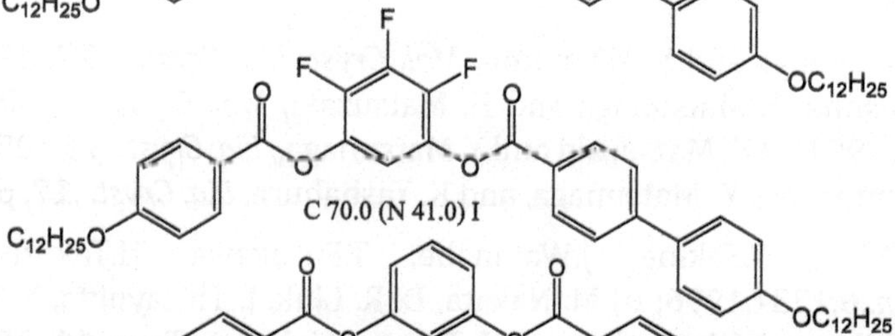

C 70.0 (N 41.0) I

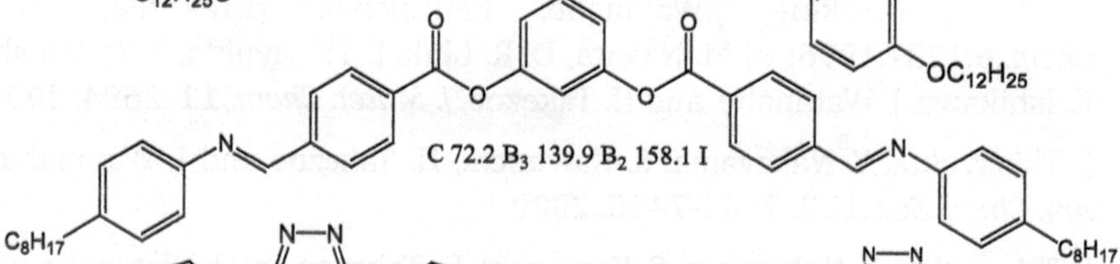

C 72.2 B_3 139.9 B_2 158.1 I

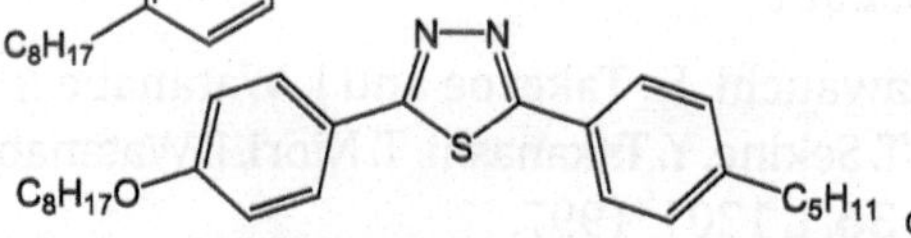

C 80.0 S_C 167.0 N 182.0 I

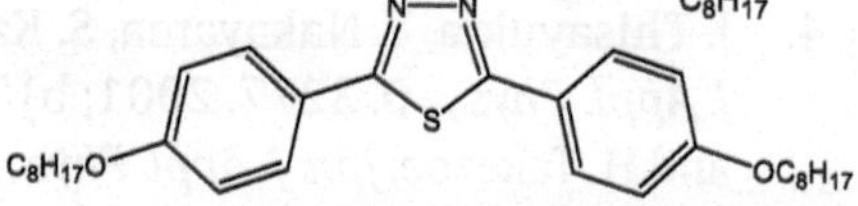

C 153.5 S_C 160.5 N 167.0 I

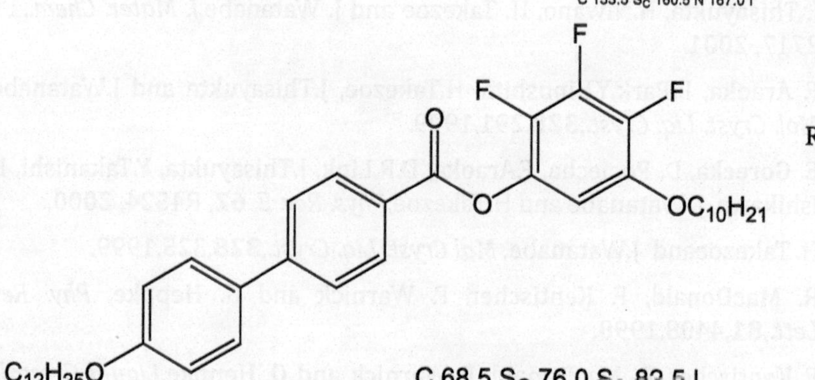

Ref. 72

C 68.5 S_C 76.0 S_A 82.5 I

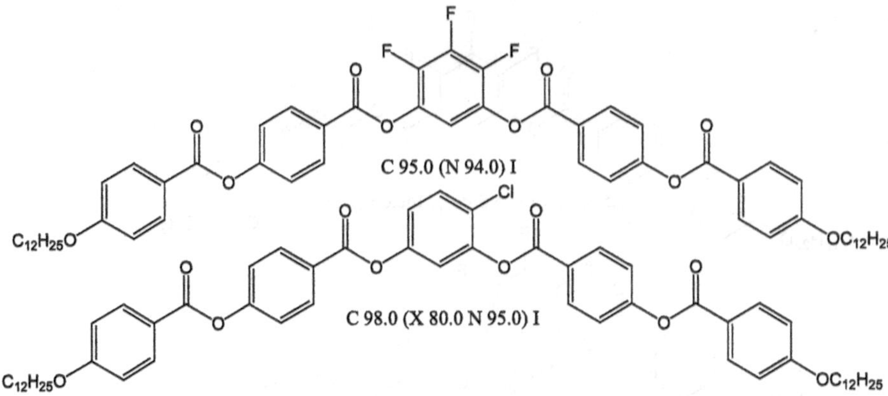

Figure 2.17: Structures of some banana shaped liquid crystals showing nematic phase

References

1. Y. Matsunaga and S. Miyamoto, *Mol. Cryst. Liq. Cryst.*, **237**, 1993; b) M. Kuboshita, Y. Matsunaga and H. Matsuzaki, *Mol. Cryst. Liq. Cryst.*, **199**, 319, 1991; c) H. Matsuzaki and Y. Matsunaga., *Liq. Cryst.*, **14**,105, 1993; d) T. Akutagawa, Y. Matsunaga, and K. Yashahura, *Liq. Cryst.* ,**17**, 659, 1994.
2. T. Niori, T.Sekine, J.Watanabe, T.Furukawa, H.Takezoe,*J. Mat. Chem.*,**6**,1231,1996; b) M. Nakata, D. R. Link, J. Thisayukta, Y. Takanishi, K. Ishikawa, J. Watanabe, and H. Takezoe, *J. Mater. Chem.*,**11**, 2694, 2001.
3. J. Thisayukta, Y. Nakayama, S. Kawauchi, H. Takezoe and J. Watanabe,*J. Am. Chem. Soc.*,**122**, 7441-7448, 2000.
4. J. Thisayukta, Y. Nakayama, S. Kawauchi, H. Takezoe and J. Watanabe,*Jpn. J. Appl. Phys.*,**40**, 3277, 2001; b) T. Sekine, Y. Takanashi, T. Niori, J. Watanabe, and H. Takezoe, *Jpn. J. Appl. Phys.*, **36**, L 1201, 1997.
5. J. Thisayukta, H.Niwano, H. Takezoe and J. Watanabe,*J. Mater. Chem.*.**11**, 2717, 2001.
6. F. Araoka, B.Park,Y.Kinoshita, H.Takezoe, J.Thisayukta and J.Watanabe, *Mol. Cryst. Liq. Cryst.*,**328**,291,1999.
7. E. Gorecka, D. Pociecha, F.Araoka, D.R.Link, J.Thisayukta, Y.Takanishi, K. Ishikawa, J. Watanabe and H.Takezoe,*Phys. Rev. E*, **62**, R4524, 2000.
8. H. Takezoeand J.Watanabe, *Mol Cryst. Liq. Cryst.*,**328**,325,1999.
9. R. MacDonald, F. Kentischer, P. Warnick and G. Heppke, *Phy. Rev. Lett.*,**81**,4408,1998.
10. F. Kentischer, R. MacDonald, P. Warnick and G. Heppke,*Liquid Crystals*, **25**,341,1998.

11. E. A. Soto Bustamante, S. V. Yablonskii, B. I. Ostrovskii, L. A. Beresnev, L. M. Blinov and W. Haase, *Liq. Cryst.*, **21**, 829; 1996; b) L.M.Blinov,*Liq. Cryst.*,**24**,143,1998.

12. G. Pelzl, S.Dieleand W.Weissflog,*Adv. Mat.*,**11**,707, 1999 and references there in.

13. W.Weissflog, H. Nadasi, U. Dunemann, G. Pelzl, S. Diele, A. Aremin and H. Kresse, *J. Mater.Chem.*, **11**, 2748, 2001; b) U. Dunemann, M. W. Schroder, R. A. Reddy, G. Pelzl, S. Diele, W. Weissflog, *J. Mater.Chem.*, **15**, 4051, 2005.

14. D. M. Walba, E. Korblova, R. Shao, and N. A. Clark, *J. Mater. Chem.*, **11**, *2743*, 2001.

15. C.V.Yelamaggad, U. S.Hiremath, S.A.Nagamani, D.S.S.Rao and S.K.Prasad,*J. Mater. Chem.*,**11**, 1818, 2001.

16. D.R.Link, G.Natale,R.Shao, J.E.Maclennan, N.A.Clark, E. Korblova, and D.M.Walba, *Science*, **278**, 1924,1997.

17. D. M. Walba, E. Korblova, R. Shao, J. E. Maclennan, D. R. Link, M. A. Glaser, N. A. Clark, *Science*, **288**, 2181, 2000.

18. N. V. S. Rao, S. Das and M. K. Paul presented at National Liquid crystal conference 2001 submitted for publication; Saibal Das M.Sc. Dissertation submitted to Assam University 2001;b) N. V. S. Rao, R. Deb, M. K. Paul and T. Francis, *Liq. Crys*,,**36**, 977–987, 2009.

19. J. P. Bedel, J. C. Rouillon, J. P. Marcerou, M. Laguerre, H. T. Nguyenand M. F. Achard,*J. Mater. Chem.*, **12**, 2214, 2002.

20. J. C. Rouillon, J. P. Marcerou, M. Laguerre, H. T. Nguyen and M. F. Achard, *J. Mater. Chem*, **11**, 2946, 2001.

21. W. Weissflog, I. Wirth, S. Diele, G. Pelzl and H. Schmalfuss, *Liq. Cryst.*, **28**, 1603, 2001.

22. G.Heppke and G.Moro,*Science*,**27**,1872,1998.

23. D. Shen, A. Pegenau, S. Diele, I. Wirth and C. Tschierske,*J. Am. Chem. Soc.* **122**, 1593, 2000.

24. D. Shen, S. Diele, G.Pelz, I. Wirth and C. Tschierske,*J.Mater. Chem.*,**9**, 661,1999.

25. D. Shen, S. Diele, I. Wirth and C. Tschierske,*Chem. Commun.*, 2573,1998.

26. R. A. Reddy and C. Tschierske, *J. Mater. Chem.*,**16**, 907, 2006.

27. H. Takezoe, Y.Takanishi, *Jpn. J. Appl. Phys.*,***45***, 597–625, 2006.

28. E. Gorecka, D. Pociecha,F. Araoka, D. R. Link, M. Nakata, J. Thisayukta, Y. Takanishi, K. Ishikawa, J. Watanabe, H. Takezoe, *Phys. Rev. E*, ***62***, R4524-R4527, 2000.

29. M. Nakata, D. R. Link, F. Araoka, J. Thisayukta, Y. Takanishi, K. Ishikawa, J. Watanabe, H. Takezoe, *Liq.Cryst.*,***28***, 1301-1308, 2001.

30. K. Kumazawa, M. Nakata, F. Araoka, Y. Takanishi, K. Ishikawa, J. Watanabe, H. Takezoe,*J. Mater. Chem.*,**14**, 157,2004.

31. F. Araoka, J. Thisayukta, K. Ishikawa, J. Watanabe, H. Takezoe, *Phys.Rev E*, **66**, 21705, 2002.

32. S. K. Lee, S. Heo, J. G. Lee, K. Kang, K. Kumazawa, K. Nishida, Y. Shimbo, Y. Takanishi,J. Watanabe, T. Doi, T. Takahashi, and H. Takezoe, *J. Am. Chem. Soc.*,**127**, 11085, 2005.

33. C. Keith, R. A. Reddy, M. Prehm, U. Baumeister,H. Kresse, J. L. Chao, H. Hahn, H. Lang and C. Tschierske, *Chem. Eur. J.*, **13**, 2556, 2007.

34. a) C. Keith, R. A. Reddy, A. Hauser, U. Baumeister and C. Tschierske, *J. Am. Chem. Soc.*,**128**, 3051, 2006; b) C. Keith, R. A. Reddy, U.Baumeister, H. Hahn, H. Lang, C. Tschierske, *J. Mater.Chem.*, **16**, 3444, 2006.

35. R. A. Reddy, U. Baumeister, C. Keith, H. Hahn, H. Lang and C.Tschierske, *Soft Matter*, 3, 5, 2007. b) R. A. Reddy, U. Baumeister, and C. Tschierske,*Chem. Comm.*, 4236,2009.

36. J. Barbera, I. De Francisco, P. Forcen, N. Gimeno, L. Monreal, M. B. Ros and J. L. Serrano, 9th International Ferroelectric Liquid Crystal Conference, 2003; b) S. Rauch, C. Selbmann, P. Bault, H. Sawade, G. Heppke, O. Morales-Saavedra, M. Y. M. Huang and A. Jakli, *Phys. Rev. E*, **69**, 021707, 2004. c) J. Barbera, I. De Francisco, P. Forcen, N. Gimeno, L. Monreal, M. B. Ros and J. L. Serrano, 9th International Ferroelectric Liquid Crystal Conference, 2003; d) G. Dantlgraber, D. Shen, S. Diele, and C. Tschierske, *Chem. Mater.*, **14**, 1149, 2002; e) C. Keith, R. Amaranatha Reddy, M. Prehm, U. Baumeister, H. Hahn, H. Lang and C. Tschierske, presented at the International Liquid Crystal Conference, 2004.

37. W.Weissflog, H. N. S. Murthy, S. Diele and G. Pelzl, *Phil. Trans. R. Soc. A*, **364**, 2657, 2006; b) W. Weissflog, G. Naumann, B. Kosata, M. W. Schroder, A. Eremin, S. Diele, Z. Vakhovskaya, H. Kresse, R. Friedemann, S. A. Rama Krishnan and G. Pelzl, *J. Mater. Chem.*, **15**, 4328–4337, 2005.

38. K. Gomola, L. Guo, S. Dhara, Y. Shimbo, E. Gorecka, D. Pociecha, J. Mieczkowski and H. Takezoe, *J. Mater. Chem.*, **19**, 4240, 2009.

39. K. Gomola, L. Guo, E. Gorecka, D. Pociecha, J. Mieczkowski , K. Ishikawa and H. Takezoe,*Chem. Comm.* 6592, 2009.

40. J. P. Bedel, J. C. Rouillon, J. P. Marcerou, M. Laguerre, H. T. Nguyen,and M. F. Achard, *J. Mater. Chem.*, **12**, 2214, 2002; b) J. C. Rouillon, J. P. Marcerou, M. Laguerre, H.T. Nguyen,and M. F. Achard, *J. Mater. Chem.*, **12**, 2946, 2002; c) J. P. Bedel, J. C. Rouillon, J. P. Marcerou, M. Laguerre, H.T. Nguyen,and M. F. Achard, *Liq. Cryst.*,**27**, 1411, 2000; d) J. P. Bedel, J. C. Rouillon, J. P. Marcerou, M. Laguerre, M. F. Achard, and H.T. Nguyen,*Liq. Cryst.*,**27**, 103, 2000; e) J. P. Bedel, J. C. Rouillon, J. P. Marcerou, M. Laguerre, H.T. Nguyen and M. F. Achard,*Liq. Cryst.*, **28**, 1285, 2001; f) H.T. Nguyen,J. P. Bedel, J. C. Rouillon, J. P. Marcerou, and M. F. Achard, *Pramana*, **61**, 395, 2003;

g) H.T. Nguyen, J. C. Rouillon, J. P. Marcerou, J. P. Bedel, P. Barois and S. Sarmento, *Mol. Cryst. Liq. Cryst.*, **328**, 177, 1999.

41. H. Dehne, M. Potter, S. Sokolowski, W. Weissflog, S. Diele, G. Pelzl, I.Wirth, H. Kresse, H. Schmalfuss and S. Grande, *Liq. Cryst.*, **28**, 1269, 2001; b) W. Weissflog, I.Wirth , S. Diele, G. Pelzl, H. Schmalfuss, T. Schoss and A. Wurflinger, *Liq. Cryst.*, **28**, 1603, 2001; c) G. Pelzl, A. Eremin, S. Diele, H. Kresse and W.Weissflog, *J.Mater.Chem.*,**12**, 2591, 2002; d) G. Pelzl, I. Wirth, and W. Weissflog, *Liq. Cryst.*, **28**, 969, 2001;e) D. Vorlander, *Ber. Dtsch. Chem. Ges.*,**62**, 2831, 1929 and f) D. Vorlander and A. Apel, *Ber. Dtsch. Chem. Ges.*, **65**, 1101, 1932.

42. W. Weissflog, H. Nadasi, U. Dunemann, G. Pelzl, S. Diele, A. Eremin and H. Kresse, *J. Mater. Chem.*, **11**, 2748, 2001.

43. D. A. Coleman, J. Fernsler, N. Chattam, M. Nakata, Y. Takanishi, E. Korblova, D. R. Link, R. F. Shao, W. G. Jang, J. E. Maclennan, O. Mondainn-Monval, V. Boyet, W. Weissflog, G. Pelzl, L. C. Chien, J. Zasadizinski, J. Watanabe, D. M. Walba, H.Takezoe and N. A. Clark, *Science*, **301**, 1204, 2003.

44. P. Pye, J. Mieczkowski, D. Pociecha, E. Gorecka, B. Donnio and D. Guillon, *J. Mater. Chem.*, **14**, 2374, 2004; b) J. Mieczkowski, K. Gomola, J. Koseska, D. Pociecha, J. Szydlowska and E. Gorecka, *J. Mater. Chem.*,**13**, 2132, 2003.

45. V. Prasad, S.W. Kang, X. Qi and S. Kumar, *J. Mater. Chem.*, **14**, 1495, 2004, b) V. Prasad, S.W. Kang and S Kumar, *J. Mater. Chem.*, **13**, 1259, 2003.

46. a)P. A. Kumar, M.L.N. Madhumohan and V. G. K. M. Pisipati, *Liq. Cryst.*, **27**, 1533, 2000; b) G. Dantlgraber, S. Diele, and C. Tschierske, *Chem. Commun.*, 2768, 2002.

47. N.V.S Rao, M.K.Paul, I. Miyake, Y.Takanishi, K. Ishikawa and H.Takezoe, *J. Mater. Chem*,**13**, 2880-2884, 2003; b) I. Miyake, Y. Takanishi, N. V. S. Rao, M. K. Paul, K. Ishikawa and H. Takezoe, *J. Mater. Chem.*, **15**, 4688-4694, 2005.

48. S. W. Choi, Y. Kinoshita, B. Park, H. Takezoe, T. Niori and J. Watanabe, *Jpn. J. Appl Phys.*, **37**, 3408, 1998; b) J. Watanabe, T. Niori, S. W. Choi, Y. Takanishi, and H. Takezoe, *Jpn. J. Appl Phys*, **37**, L401, 1998.

49. S. W. Choi, M. Zennyoji, Y. Takanishi, H. Takezoe, T. Niori and J. Watanabe, *Mol. Cryst. Liq. Cryst.*, **328**, 185, 1999; b) J. Watanabe, T. Izumi, T. Niori, M.Zennyoji, Y.Takanishi, and H.Takezoe, *Mol. Cryst. Liq. Cryst.*, **346**, 77, 2000.

50. W. Weissflog, Ch. Lischka, K. Pelz, I. Wirth, S. Diele and G. Pelzl, *Mol. Cryst. Liq. Cryst.*, **364**, 279, 2001; b) W. Weissflog, Ch. Lischka, S. Diele, I. Wirth and G. Pelzl,*Liq. Cryst.*, **27**, 43, 2000.

51. V. Prasad, D. S. S. Rao, and S. K. Prasad, *Liq. Cryst.*, **27**, 585, 2000; b) ibid. **28**, 761, 2001.

52. H. N. S. Murthy and B. K. Sadashiva, *Liq. Cryst.*, **29**, 1223, 2002.

53. R. A. Reddy and B. K. Sadashiva, *Liq. Cryst.*, **30**, 1031, 2003.

54. R. A. Reddy and B. K. Sadashiva, *J. Mater. Chem.*, **12**, 2627, 2002.

■ R. A. Reddy and B. K. Sadashiva, Liq. Cryst., 29, 1365, 2002; b) ibid.30, 273, 2003.

55. B. K. Sadashiva, H. N. S. Murthy and S. Dhara ,*Liq. Cryst.*, **28**, 483, 2001.

56. S. Shubasree, B. K. Sadashiva and S. Dhara ,*Liq. Cryst.*, **29**, 789, 2002.

57. R. Pratibha, N. V. Madhususana and B. K. Sadashiva ,*Science* , **288**, 2185, 2000.

58. B. K. Sadashiva, R. A. Reddy, R. Pratibha and N. V. Madhusudana,*J. Mater. Chem.*, **12**, 943, 2001;b) B. K. Sadashiva, R. A. Reddy, R. Pratibha and N. V. Madhusudana,*Chem. Commun.*, 2140, 2001; and c) R. A. Reddy, B. Sadashiva and S. Dhara, *Chem. Commun.*, 1972, 2002.

59. R. A. Reddy and B. K. Sadashiva, *Liq. Cryst.*, **27**, 1613, 2000; b) R. A. Reddy and B. K. Sadashiva, *J. Mater. Chem.*, **14**, 1936, 2004; c) R. A. Reddy, V. A. Raghunathan and B. K. Sadashiva, *Chem. Mater.*, **7**, 274, 2005; d) R. A. Reddy, B. K. Sadashiva and V. A. Raghunathan, *Chem. Mater.*, **16**, 4050, 2005.

60. H. T. Nguyen, J. Rouillon, J. P.Marcerou, J. P.Bedel, P.Barois, S. Armento, *Mol. Cryst. Liq. Cryst.*, **328**, 177, 1999.

61. J. Svoboda, V. Novotna, V. Kozmık, M. Glogarova, W. Weissflog, S. Diele and G. Pelzl, *J. Mater. Chem.*, **13**, 2104, 2003.

62. S. Kang, Y. Saito, N. Watanabe, M. Tokita, Y. Takanishi, H. Takezoe and J. Watanabe, *J. Phys. Chem. B*,**110**, 5205, 2006.

63. F. C. Yu and L. J. Yu, *Chem Mater.*, **18**, 5410, 2006.

64. J. H. Wild, K. Bartle, M. O'Neill, S. M. Kelly and R. P. Tuffin, *Liq. Cryst.*, **33**, 635, 2006.

65. S. Sokolowski, H. Dehne, W. Weissflog, unpublished.

66. T. Masuda and Y. Matsunaga, *Bull. Chem. Soc. Jpn.*, **64**, 2192, 1991.

67. M. Hird, Y. Raoul, J. W. Goodby and H. Gleeson, *Ferroelectrics*, **309**, 95, 2004.

68. M. Hird, Banana Liquid Crystals – Chirality and Polarity, Workshop, Boulder, US, 21–24.8. 2002, oral lecture.

69. M. Hird, J. W. Goodby, P. Hindmarsh, R. A. Lewis, and K. J. Toyne, *Ferroelectrics*, **276**, 219, 2002; b) T. Geelhaar, *Ferroelectrics*, **85**, 329, 1988.

70. M. Hird, J. W. Goodby, N. Gough and K. J. Toyne, *J. Mater. Chem.*,**11**, 2732 2001.

71. M. Hird and K. J. Toyne, *Mol. Cryst. Liq. Cryst.*,323, 1, 1998.

CHAPTER 3

Synthesis and Methodologies

We describe the synthetic steps and procedure for efficient and elegant route towards successful synthesis of the desired compounds. Chemical characterisation of the compounds along with the spectral dataare presented. This chapter also gives a brief outline about the experimental setup with working principle and method involved forstudying the phase behavior, electro-optic behavior and other physic-chemical properties of liquid crystalline materials.

3. Experimental: Synthesis, Techniques and Methods

3.1. Synthesis:

The target compounds are designed and the syntheses of different compounds at intermediate stages were carried out following the procedures well documented in literature [1-3]. The different reagents and reaction conditions are modified wherever they are necessary, synthesis was carried out on a trial basis and then the trial procedure was followed to get the maximum yield of the desired compounds in pure form. The compounds are recrystallized from suitable solvents or purified by column chromatography. The synthetic procedures for the intermediate as well as target compounds are depicted bellow. All the solvents are purified following standard procedures well documented in the literature [4-6]. Silica gel (60-120 mesh) from Acme synthetic chemicals was used for chromatographic separation. Silica gel G [E-Merck (India)] was used for TLC. Benzene, Chloroform, Ethyl acetate and Petroleum ether (refers to the fraction boiling between 60°C and 80°C) were used as eluents for chromatographic separation.

3.2. Commercially available compounds

The following commercial compounds were procured from Frinton Chemicals, (U. S. A.), Tokyo Kasei Chemicals (Japan), Lancaster synthesis (U.K.), Aldrich chemicals (India) and Avocado Chemicals (U.K.).

1

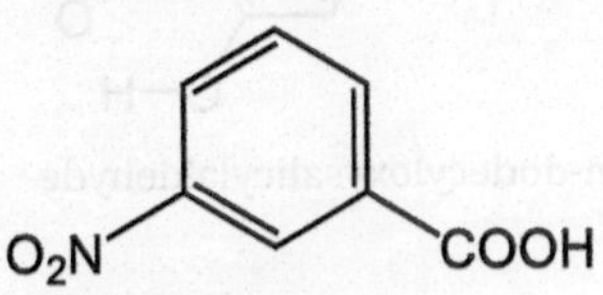

3-nitrobenzoicacid

2

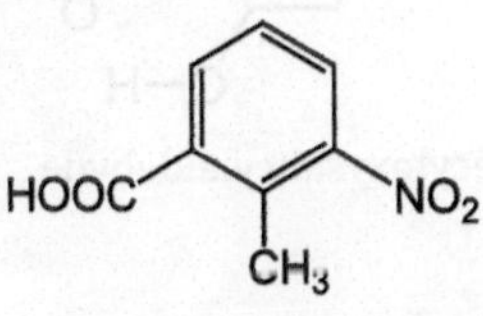

2-methyl-3-nitrobenzoicacid

3)

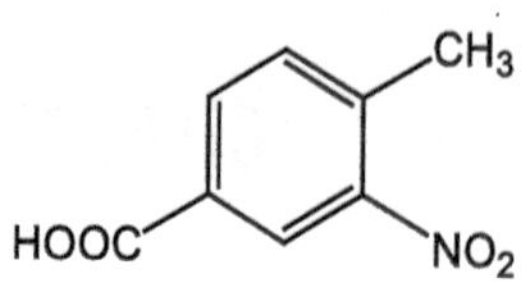

4-methyl-3-nitrobenzoicacid

4}

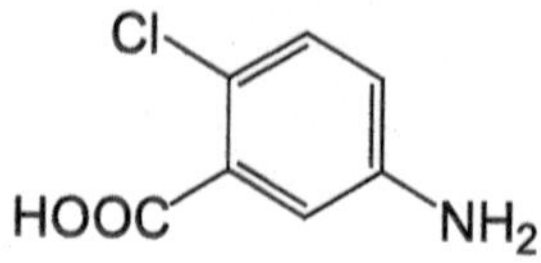

2-chloro-5-nitrobenzoicacid

5)

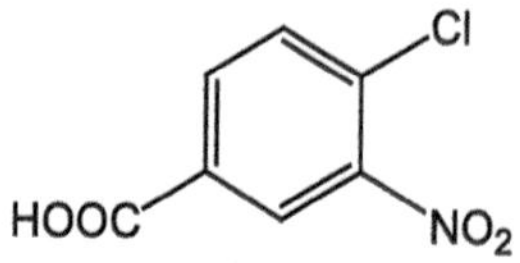

4-chloro-3-nitrobenzoicacid

6)

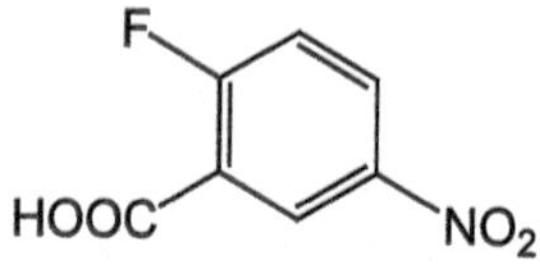

2-fluoro-5-nitrobenzoicacid

7)

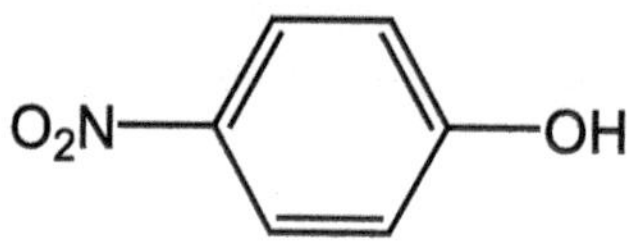

4-nitrophenol

The following intermediate compounds and final products had been synthesized in the laboratory and characterized for their molecular and material characteristics during the last few years and are furnished below.

3.3. Intermediate compounds:

1)

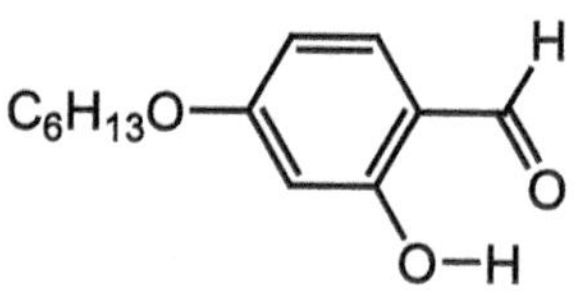

4-n-hexyloxysalicylaldehyde

2)

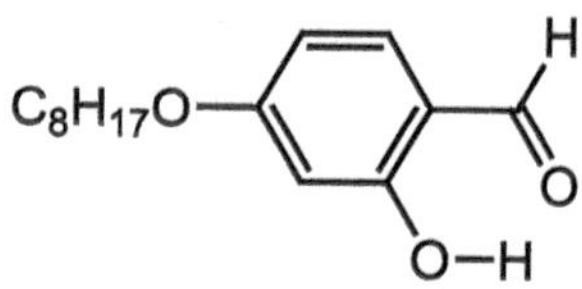

4-n-octyloxysalicylaldehyde

3)

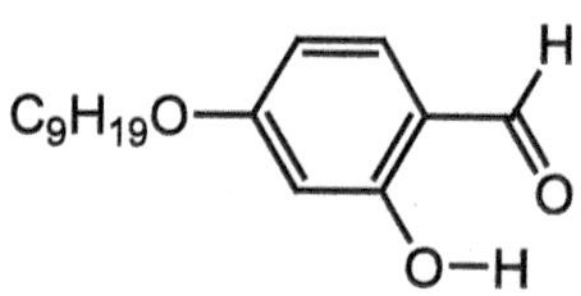

4-n-nonyloxysalicylaldehyde

4)

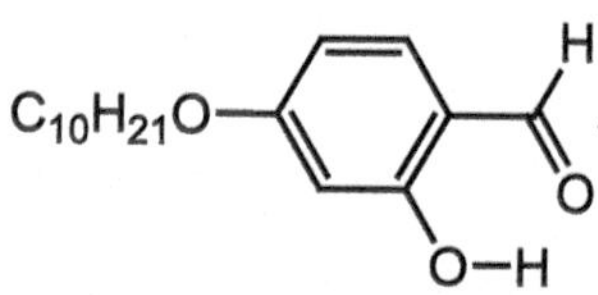

4-n-decyloxysalicylaldehyde

5)

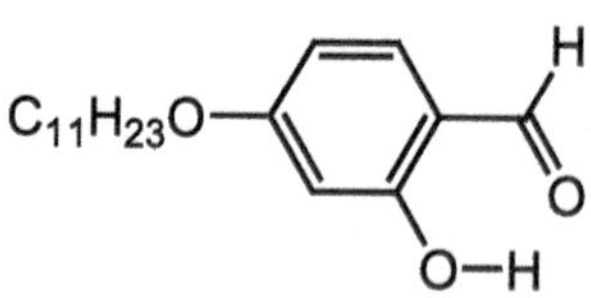

4-n-undecyloxysalicylaldehyde

6)

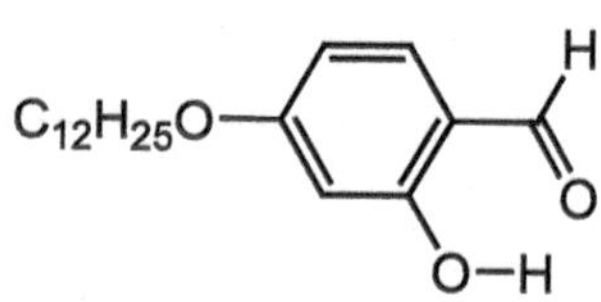

4-n-dodecyloxysalicylaldehyde

7) 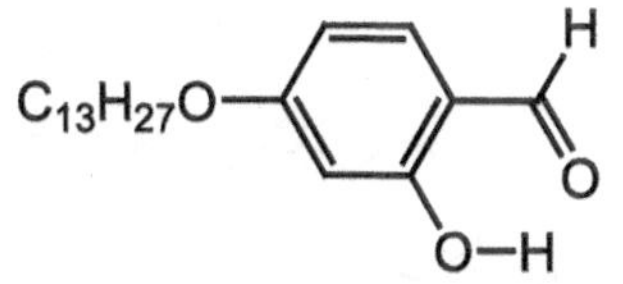

4-n-tridecyloxysalicylaldehyde

8) 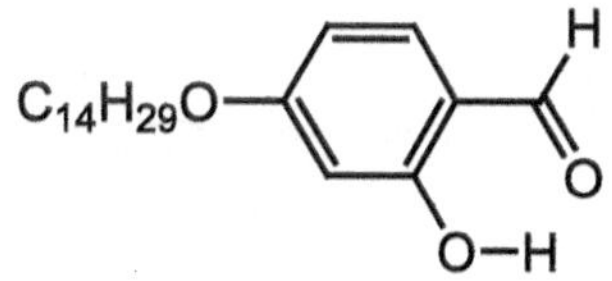

4-n-tetradecyloxysalicylaldehyde

9) 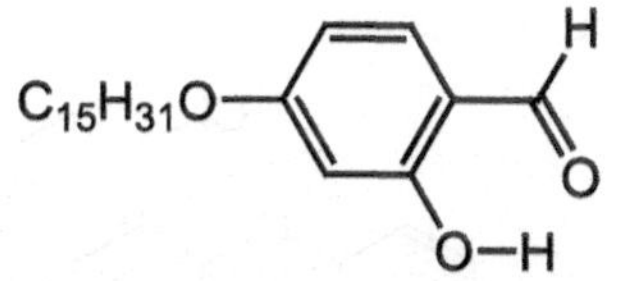

4-n-pentadecyloxysalicylaldehyde

10) 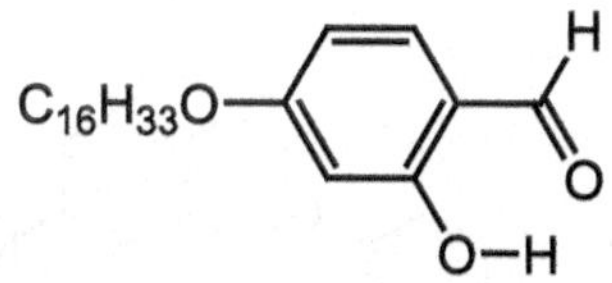

4-n-hexadecyloxysalicylaldehyde

11) 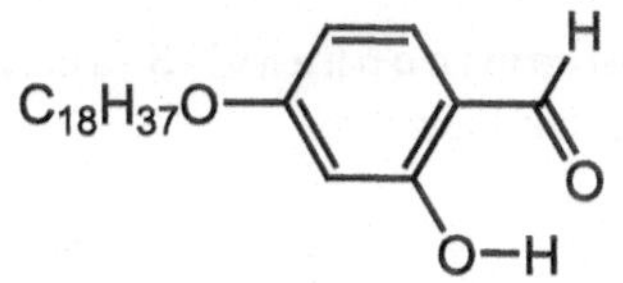

4-n-octadecyloxysalicylaldehyde,

12) 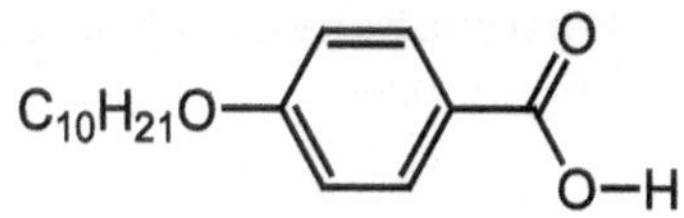

4-n-decyloxybenzoic acid

13) 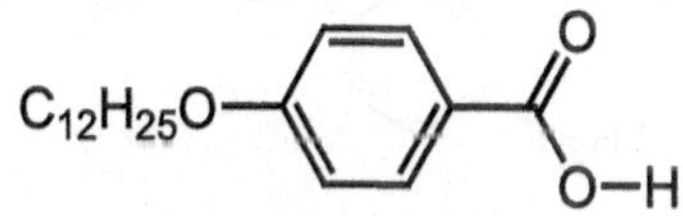

4-n-dodecyloxybenzoic acid

14) 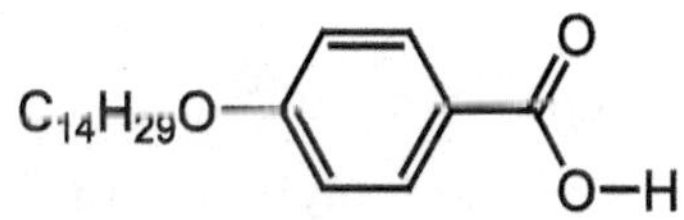

4-n-tetradecyloxybenzoic acid

15) 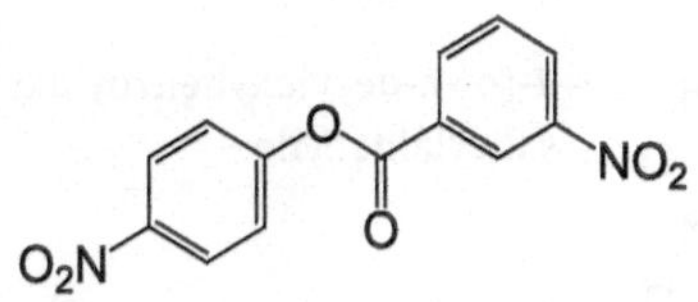

4-nitrophenyl-3-nitrobenzoate

16) 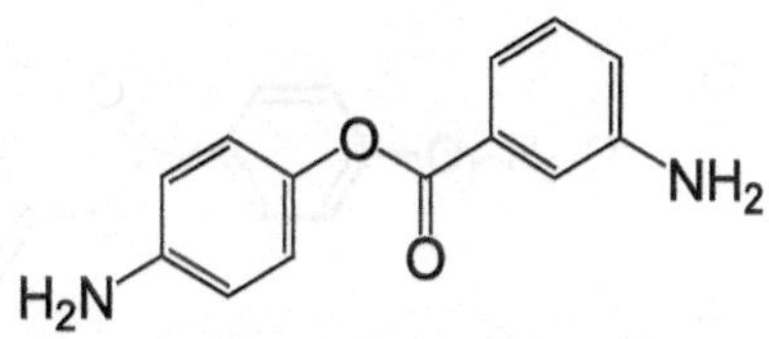

4-aminophenyl-3-aminobenzoate

17) 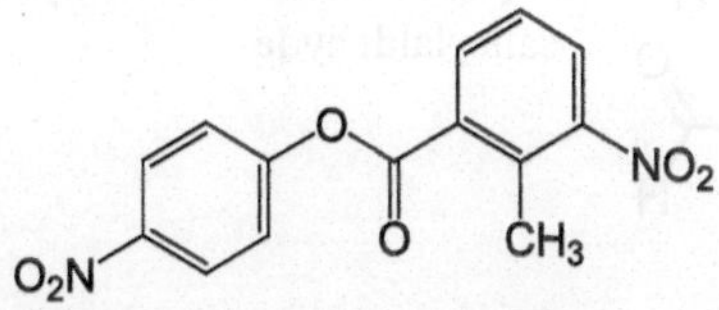

4-nitrophenyl-2-methyl-3-nitrobenzoate

18) 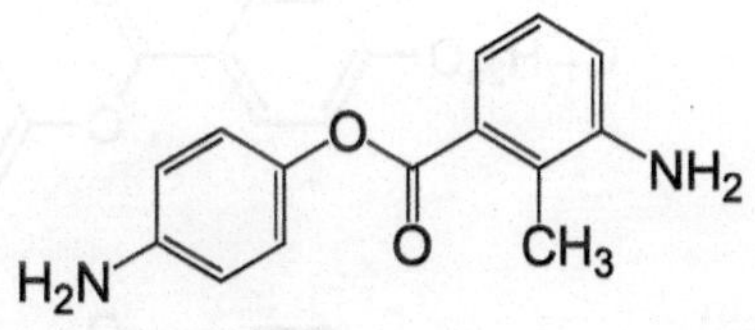

4-aminophenyl-3-amino-2-methylbenzoate

19) 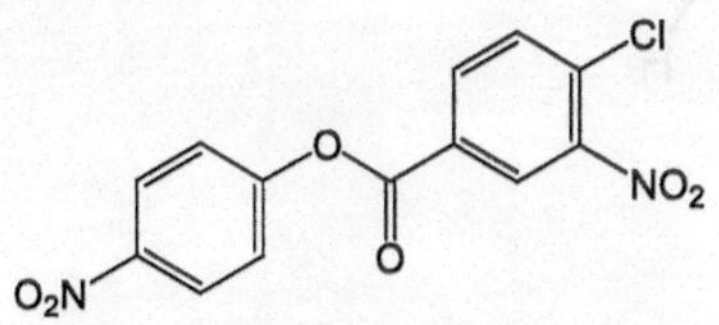

4-nitrophenyl-4-chloro-3-nitrobenzoate

20) 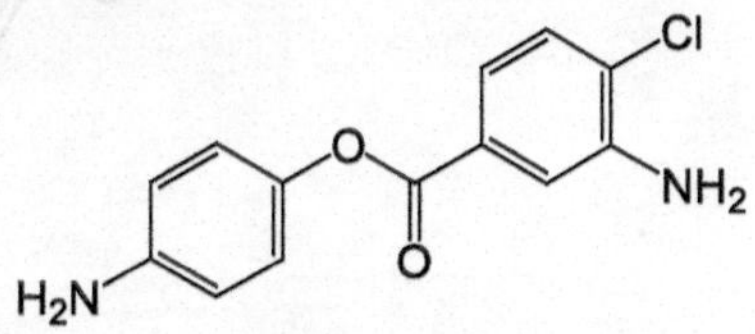

4/-aminophenyl-3-amino-4-chlorobenzoate

21)

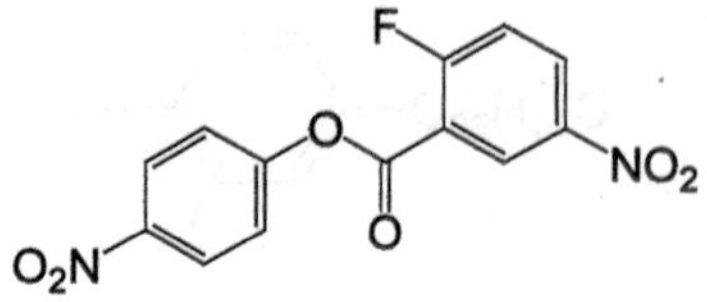

4-nitrophenyl-2-fluoro-5-nitrobenzoate

22)

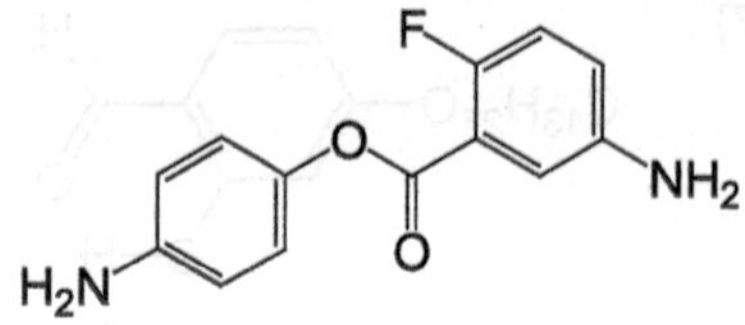

4′-aminophenyl-5-amino-2-fluorobenzoate

23)

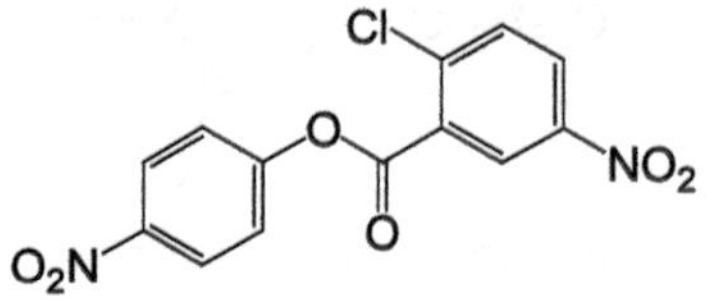

4-nitrophenyl-2-chloro-5-nitrobenzoate

26)

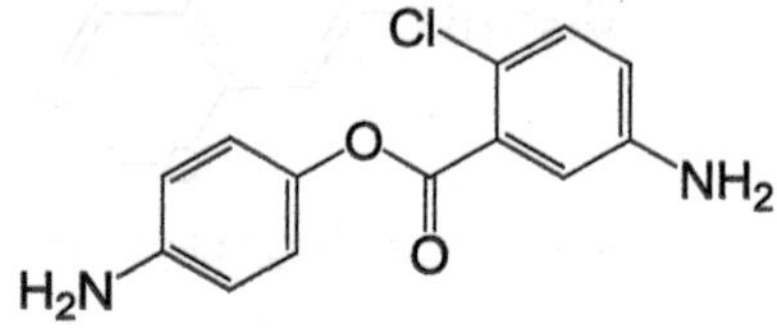

4′-aminophenyl-5-amino-2-chlorobenzoate

25)

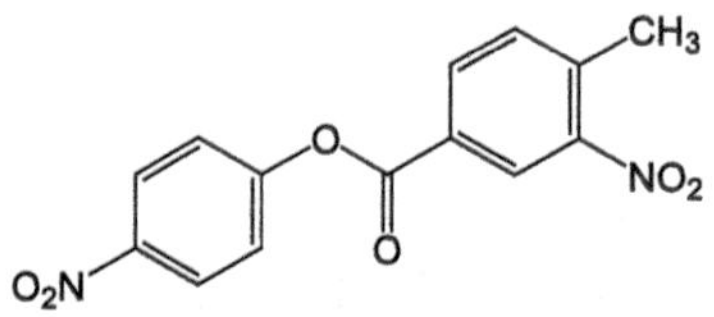

4-nitrophenyl-4-methyl-3-nitrobenzoate

26)

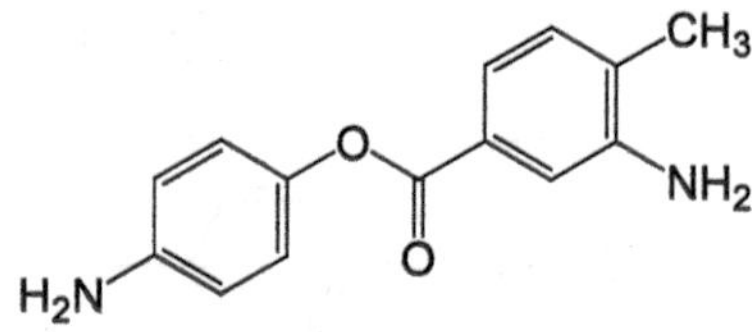

4′-aminophenyl-3-amino-4-methylbenzoate

27)

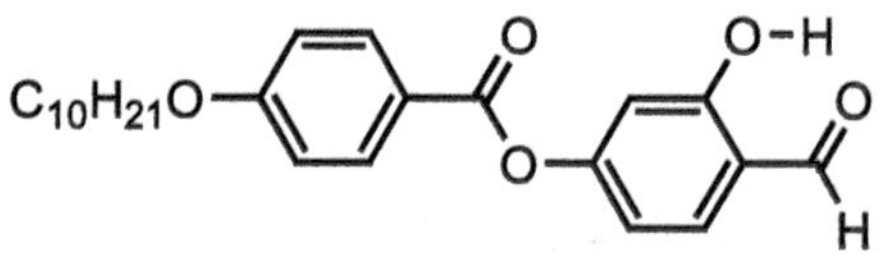

4-(4′-n-decyloxybenzoyloxy) salicylaldehyde

28)

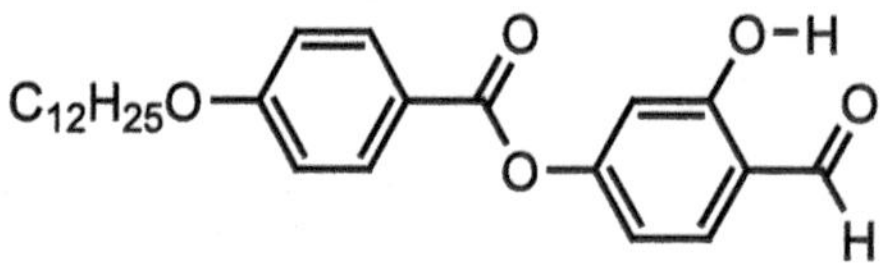

4-(4′-n-dodecyloxybenzoyloxy) salicylaldehyde

29)

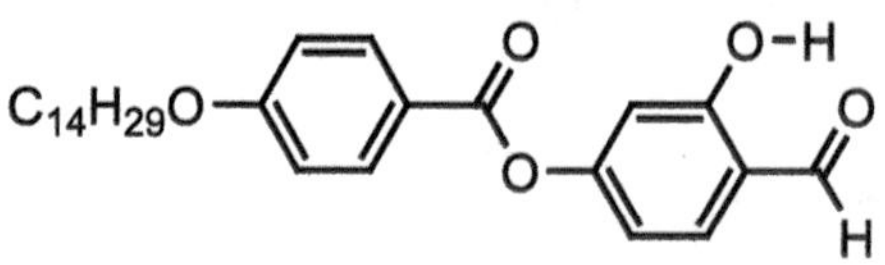

4-(4′-n-tetradecyloxybenzoyloxy) salicylaldehyde

3.4. Final Products

1

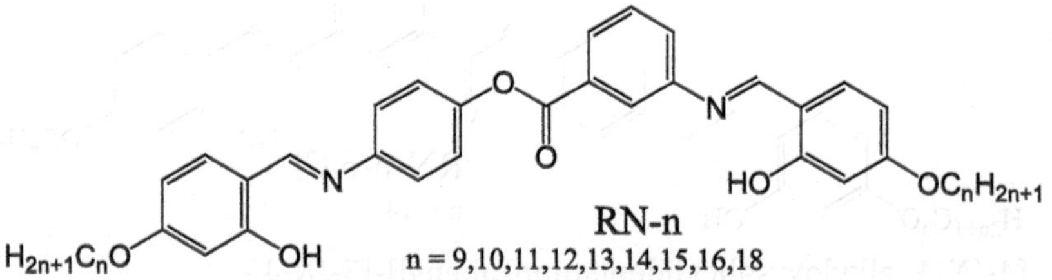

RN-n

n = 9,10,11,12,13,14,15,16,18

[4-(N-4′-alkyloxysalicylidenemino)phenyl]-[3-(N-4′-alkyloxysalicylidene-amino)benzoate]

2

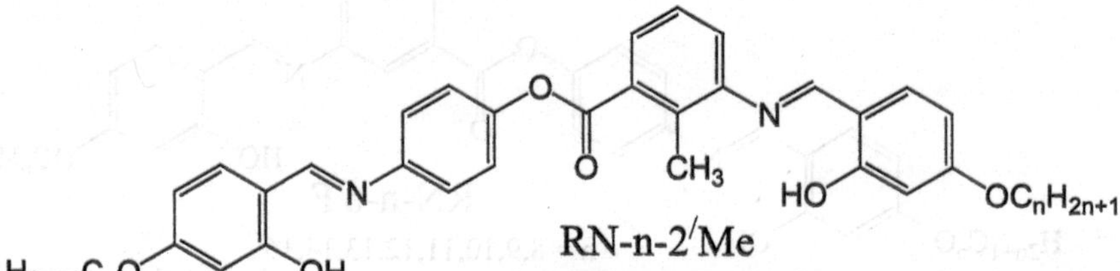

RN-n-2′Me

n = 6, 8, 10, 12, 14, 16, 18

[4-(N-4′-alkyloxysalicylidenemino)phenyl]-[3-(N-4′-alkyloxysalicylidene-amino) 2-methylbenzoate]

3

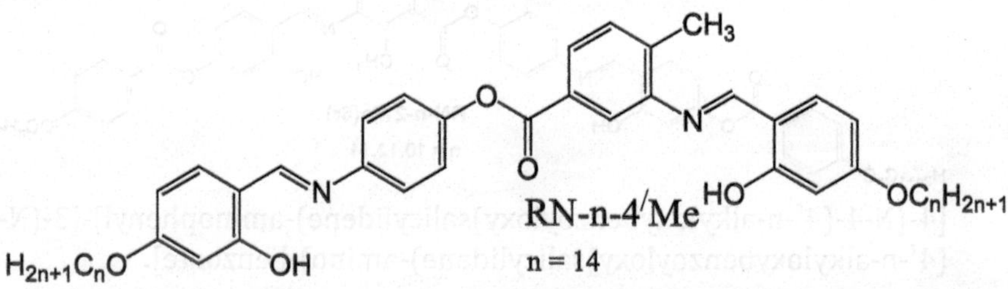

RN-n-4′Me

n = 14

[4-(N-4′-alkyloxysalicylidenemino)phenyl]-[3-(N-4′-alkyloxysalicylidene-amino)-4-methylbenzoate]

4

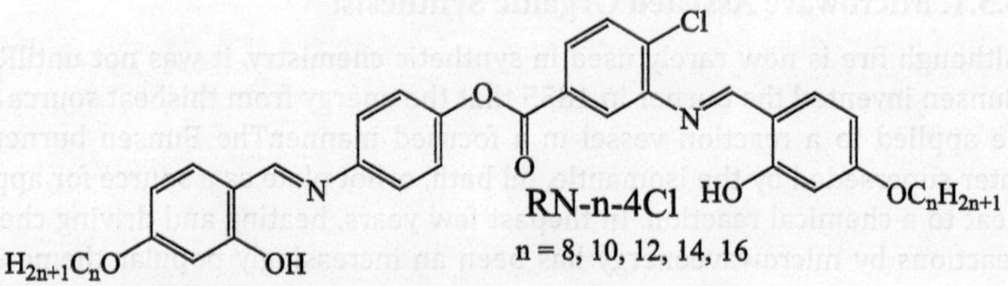

RN-n-4Cl

n = 8, 10, 12, 14, 16

[4-(N-4′-alkyloxysalicylidenemino)phenyl]-[3-(N-4′-alkyloxysalicylidene-amino)-4-chlorobenzoate]

5

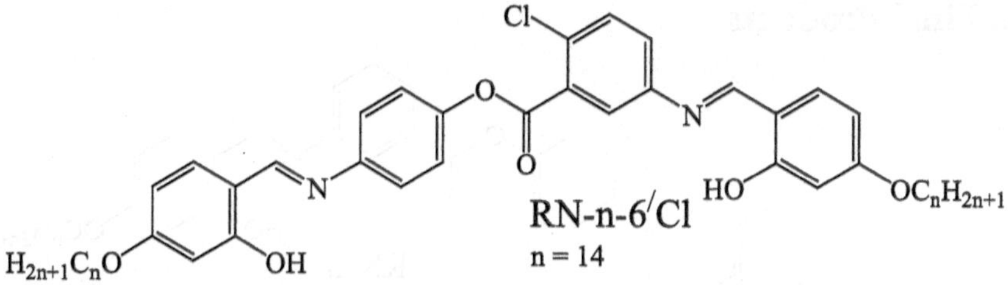

[4-(N-4′-alkyloxysalicylidenemino)phenyl]-[5-(N-4′-alkyloxysalicylidene-amino)-2-chlorobenzoate]

6

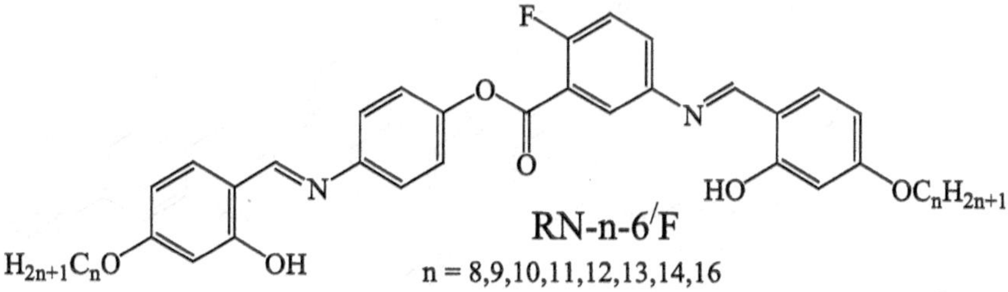

[4-(N-4′-alkyloxysalicylidenemino)phenyl]-[5-(N-4′-alkyloxysalicylidene-amino)-2-fluorobenzoate]

7

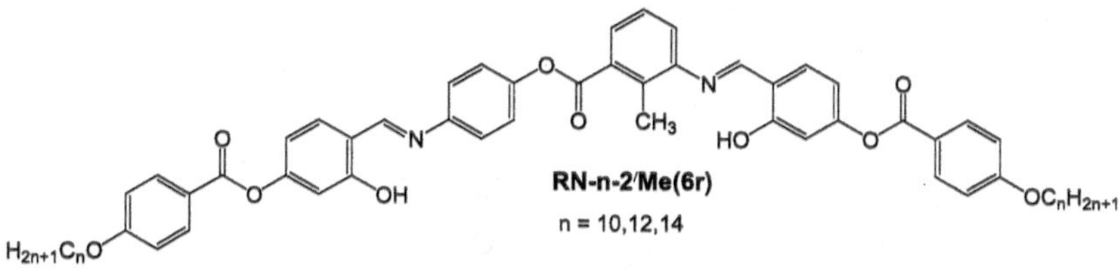

[4-(N-4-(4′-n-alkyloxybenzoyloxy)salicylidene)-aminophenyl]-[3-(N-4-(4′-n-alkyloxybenzoyloxy)salicylidene)-amino)-benzoate].

3.5. Synthesis of intermediates:

3.5.1. Microwave Assisted Organic Synthesis:

Although fire is now rarely used in synthetic chemistry, it was not untilRobert Bunsen invented the burner in 1855 that the energy from thisheat source could be applied to a reaction vessel in a focused manner.The Bunsen burner was later superseded by the isomantle, oil bath, orhot plate as a source for applying heat to a chemical reaction. In thepast few years, heating and driving chemical reactions by microwaveenergy has been an increasingly popular theme in the scientificcommunity. This nonclassical heating technique is slowlymoving froma laboratory curiosity to an established technique that is heavily used inboth academia and industry. The efficiency of "microwave flashheating" in dramatically reducing reaction times (from days and hoursto minutes and seconds) is just one of the many advantages.

High-speed synthesis with microwaves has attracted a considerable amount of attention in recent years.[7] More than 2000 articles have been published in the area of microwave assisted organic synthesis (MAOS) since the first reports on the use of microwave heating to accelerate organic chemical transformations by the groups of Gedye and Giguere/ Majetich in 1986.[8,9] The initial slow uptake of the technology in the late 1980s and early 1990s has been attributed to its lack of controllability and reproducibility, coupled with a general lack of understanding of the basics of microwave dielectric heating. The risks associated with the flammability of organic solvents in a microwave field and the lack of available systems for adequate temperature and pressure controls were major concerns. Although most of the early pioneering experiments in MAOS were performed in domestic, sometimes modified, kitchen microwave ovens, the current trend is to use dedicated instruments which have only become available in the last few years for chemical synthesis. The number of publications related to MAOS has therefore increased dramatically since the late 1990s to a point where it might be assumed that, in a few years, most chemists will probably use microwave energy to heat chemical reactions on a laboratory scale. Not only is direct microwave heating able to reduce chemical reaction times from hours to minutes, but it is also known to reduce side reactions, increase yields, and improve reproducibility. Therefore, many academic and industrial research groups are already using MAOS as a forefront technology for rapid optimization of reactions, for the efficient synthesis of new chemical entities, and for discovering and probing new chemical reactivity. A large number of review articles [10–19] and several books [20–22] provide extensive coverage of the subject.

Microwave Theory:

Microwave irradiation is electromagnetic irradiation in the frequency range of 0.3 to 300 GHz. All domestic "kitchen" microwave ovens and all dedicated microwave reactors for chemical synthesis operate at a frequency of 2.45 GHz (which corresponds to a wavelength of 12.24 cm) to avoid interference with telecommunication and cellular phone frequencies. The energy of the microwave photon in this frequency region (0.0016 eV) is too low to break chemical bonds and is also lower than the energy of Brownian motion. It is therefore clear that microwaves cannot induce chemical reactions.[14,23–25] Microwave-enhanced chemistry is based on the efficient heating of materials by "microwave dielectric heating" effects. This phenomenon is dependent on the ability of a specific material (solvent or reagent) to absorb microwave energy and convert it into heat. The electric component [26] of an electromagnetic field causes heating by two main mechanisms: dipolar polarization and ionic conduction. Irradiation of the sample at microwave frequencies results in the dipoles or ions aligning in the applied electric field. As the applied field oscillates, the dipole or ion field attempts to realign itself with the alternating electric field and, in the process, energy is lost in the form of heat through molecular friction and dielectric loss. The amount of heat generated by this process is directly related to the ability of the matrix to align itself with the frequency of the applied field. If the dipole does not have enough time to realign, or reorients too quickly with the applied field, no heating occurs. The allocated frequency of 2.45 GHz used in all commercial systems lies between these two extremes and gives the molecular dipole time to align in the field, but not to follow the alternating field precisely.[24, 25] The heating characteristics of a particular material (for example, a solvent) under microwave irradiation conditions are dependent on its dielectric properties. The ability of a specific substance to convert electromagnetic energy into heat at a given frequency and temperature is determined by the so-called loss factor tanδ. This loss factor is expressed as the quotient $\tan \delta = \varepsilon'' / \varepsilon'$, where ε'' is the dielectric loss, which is indicative of the efficiency with which electromagnetic radiation is converted into heat, and ε' is the dielectric constant describing the ability of molecules to be polarized by the electric field. A reaction medium with a high tanδ value is required for efficient absorption and, consequently, for rapid heating. The loss factors for some common organic solvents are summarized in Table 1. In general, solvents can be classified as high (tanδ>0.5), medium (tanδ 0.1–0.5), and low microwave absorbing (tanδ<0.1).

Table 3.1: Loss factor (tanδ) of different solvents[21].

Solvent	tanδ	Solvent	tanδ	Solvent	tanδ
ethylene glycol	1.350	2-butanol	0.447	chloroform	0.091
ethanol	0.941	1,2-dichlorobenzene	0.280	acetonitrile	0.062
DMSO	0.825	NMP	0.275	ethyl acetate	0.059
2-propanol	0.799	acetic acid	0.174	acetone	0.054
formic acid	0.722	DMF	0.161	THF	0.047
methanol	0.659	1,2-dichloroethane	0.127	dichloromethane	0.042
nitrobenzene	0.589	water	0.123	tolune	0.040
1-butanol	0.571	chlorobenzene	0.101	hexane	0.020

In the present study, microwave assisted synthesis were carried out in a kitchen convection microwave oven, LG MC 7648WH 26 Ltr .(360W).

4-n-decyloxybenzoic acid:

The synthesis of 4-n-decyloxybenzoic acid is carried out using microwave wave assisted synthesis technique. The starting materials 4-hydroxyethylbenzoate (0.1mol, 16.6g) is dissolved in 10 ml of NMP in a 250 ml of round bottom flask. To this n-bromodecane(0.11mol, 24.4ml) and $CsCO_3$ (0.11mol, 2.1g) is added and mixed well. The mixture is then irradiated to microwave heating for 60 sec (4x15 sec, 360 W). The reaction mixture is then cooled to room temperature and 100 ml ethanol is added. The undissolved $CsCO_3$ was filtered off. To the filtrate, 200ml of water and 8 g of NaOH is also added and refluxed for 10 hours. The reaction mixture was cooled to room temperature. After that the solution is taken in a 1000ml beaker and this solution is acidified with HCl (till pH = 2) with constant stirring. The solid white precipitate 4-n-decyloxy benzoic acid separates out from the reaction mixture and the separated white precipitate was filtered and washed with excess of water. The resulting solid was recrystallized from absolute ethanol using animal charcoal. The solid is dried to yield a pure white solid. Yield: 24g (88%).

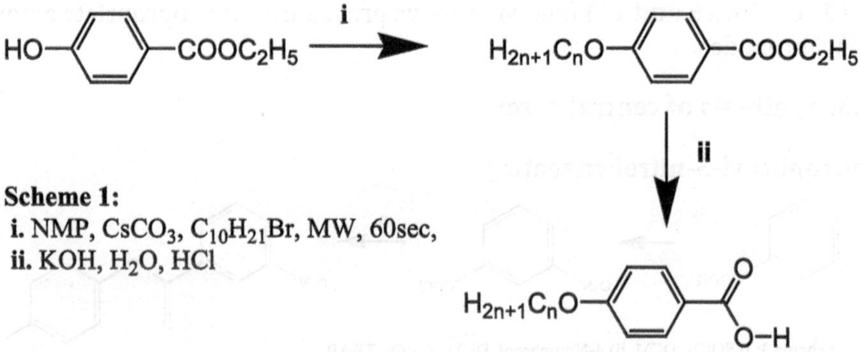

Scheme 1:
i. NMP, $CsCO_3$, $C_{10}H_{21}Br$, MW, 60sec,
ii. KOH, H_2O, HCl

The other 4-n-alkyloxysalicyladehydes were prepared **(n = 12 and 14)** following above procedure with appropriate amount of alkyl bromides.

3.5.2. Synthesis of alkoxysalicylaldehyde:

Alkylation of 2,4-dihydroxy benzaldehyde using MAOS technique yield aldehyde with both hydroxyl group alkylated. Microwave dielectric heating generates sufficient thermal energy to alkylate both the ortho (H-bonded) and para hydroxyl group of 2,4-dihydroxy benzaldehyde. Therefore to synthesise only para alkylated salicylaldehyde, microwave heating technique was not used and the reaction was carried out in conventional heating in isomental.

4-n-octyloxysalicylaldehyde:

Scheme 2:

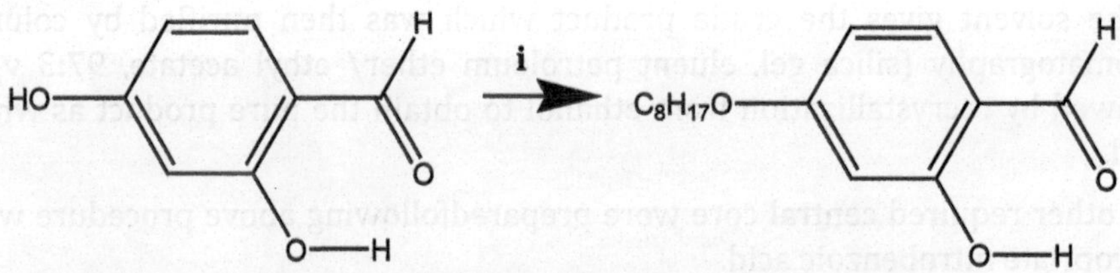

Scheme 2: i. acetone(dry), $NaHCO_3$, $C_8H_{17}Br$, KI, reflux,48 h

The synthesis of 4-alkoxy-2-hydroxy benzaldehyde, (**Scheme 2**) viz., mono alkylation was performed by using a modification of the literature procedure to improve the product yield. 2,4-dihydroxybenzaldehyde (10 g, 72.4 mmol), 1-bromooctane (16.6 ml, 75 mmol), $NaHCO_3$ (6.30 g, 75 mmol) and KI (catalytic amount) were mixed in dry acetone (250ml) and then the mixture was refluxed for 48 hours. It was then filtered hot to remove the insoluble solid. To neutralize the warm solution dilute HCl was added, which was then extracted twice with

$CHCl_3$ (100ml). The combined extracts were concentrated to give a purple solid. The product was purified by column chromatography using silica gel (60-120 mesh) eluting with a mixture of chloroform and hexane (V/V; 1/1) followed by evaporation of solvent. It gave the product as a pale yellow liquid.

The other required 4-n-alkyloxysalicyladehydes were prepared **(n = 9, 10, 11, 12, 13, 14, 15, 16 and 18)** following above procedure with appropriate amount of alkyl bromides.

3.5.3. Synthesis of central core:

4/-nitrophenyl-3-nitrobenzoate:

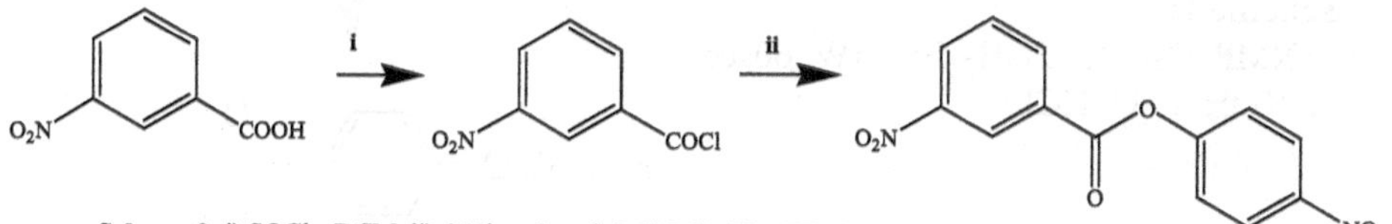

Scheme 3: i) $SOCl_2$, DCM; ii) 4-Nitrophenol, DCM, K_2CO_3, TBAB

3-nitrobenzoic acid (2.50g; 15mmol) was dissolved in dichloromethane in a two neck round bottomed flask with a teflon coated magnetic stirrer and the reaction flask was flushed with N_2, sealed with a rubber septum and cooled in an ice bath. Thionylchloride (2.0 ml, 16.5 mmol) was added drop wise slowly to the cooled reaction mixture. The ice bath was removed and the reaction mixture was vigorously stirred and refluxed for 1h. The solvent and excess thionyl chloride was evaporated under reduced pressure and the resulting compound was dried under vacuum. To the resulting acyl chloride dissolved in dichloromethane (30ml), an aqueous solution of 4-nitro phenol (2.08g; 15mmol) and K_2CO_3 (4.14g; 30mmol) were added. The resulting solution was vigorously stirred for 24 h after adding a catalytic amount of tetra butyl ammonium bromide. After the stirring was complete, the organic layer was separated, washed several times with the alkaline solution and water and then dried over sodium sulphate. Evaporation of the solvent gives the crude product which was then purified by column chromatography (silica gel, eluent petroleum ether/ ethyl acetate, 97:3 v/v) followed by recrystallization from ethanol to obtain the pure product as white solid.

The other required central core were preparedfollowing above procedure with appropriate nitrobenzoic acid.

4/-Aminophenyl-3-aminobenzoate:

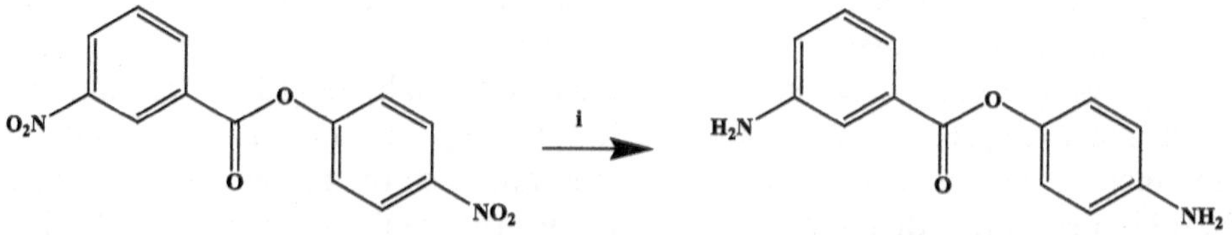

Scheme 9: i) Pd/C (10%), H_2, ethylacetate

4/-nitrophenyl-3-nitrobenzoate (2.88 g, 10 mmol) was dissolved in ethyl acetate in a two neck round bottom flask to it 10% Pd-C (0.28g). The reaction mixture is stirred for 15hrs under balloon pressure filled with hydrogen gas. On completion of the reaction the crude product was collected by evaporating the solvent under reduced pressure. The crude product was then purified by column chromatography (silica gel 60-120 mesh, using dichloromethane/ethanol, (99.5:0.5::V/V) as eluent to give pure product. Similar procedure was applied for reduction of all other di-nitro central core.

3.5.4. Synthesis of ester salicylaldehyde:

4-(4'-n-decyloxybenzoyloxy)salicylaldehyde:

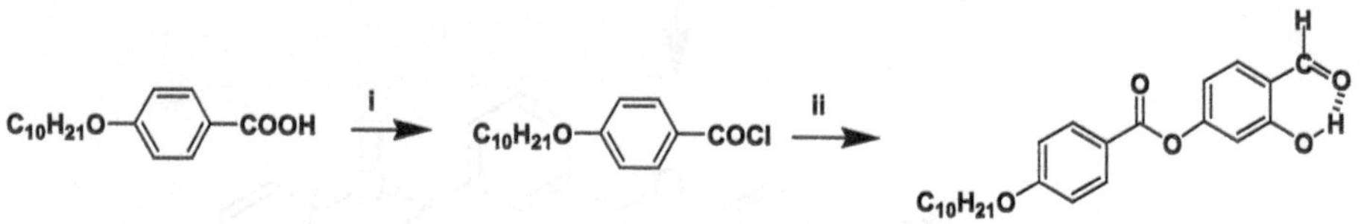

Scheme 15: i). $SOCl_2$, DCM, DMF; ii) 2,4-dihydroxybenzaldehyde, DCM, TEA

In a two-neck round bottom flask with a teflon-coated magnetic stirrer, 4-n-octyloxybenzoic acid (3.75g; 15mmol) in CH_2Cl_2, and a catalytic amount of DMF was added. The reaction flask was flushed with N_2, sealed with a rubber septum, and cooled in an ice bath. An excess of thionyl chloride, $SOCl_2$ (8ml, 2.0equiv) was added drop wise to the cooled reaction mixture. The ice bath was removed and the reaction mixture was stirred for 1 h. The solvent was evaporated under reduced pressure and the resulting compound was dried under vacuum. The product was used for the next step without further purification.To the resulting acyl chloride which was dissolved in dichloromethane (30 ml) and in presence of triethylamine (2ml) a solution of 2,4-dihydroxybenzaldehyde (2.07 g, 15 mmol) in dry dichloromethane was added drop wise with constant stirring. The resultant mixture was then vigorously stirred for 5-6 h maintaining the temperature at 10-20^{0}C. The organic layer was dried over sodium sulphate. Evaporation of the solvent gave the crude product which was then purified by column chromatography (silica gel, eluent petroleum ether/ethyl acetate, 95:5, V/V) and the solvent was evaporated to give pure product. Further purification by recrystallization from ethanol afforded the pure product as white solid.

The other required 4-n-alkoxybenzyloxysalicylaldehydes were prepared **(n = 12, 14)** following above procedure with appropriate alkoxy benzoic acid. The respective yields of the aldehydes, spectroscopic and analytical data are presented below.

3.6. Synthesis of the target designed compounds:

3.6.1. [4-(N-4-alkyloxysalicylidenemino)phenyl]-[3-(N-4-alkyloxysalicylidene-imino)benzoate]: (RN-n)

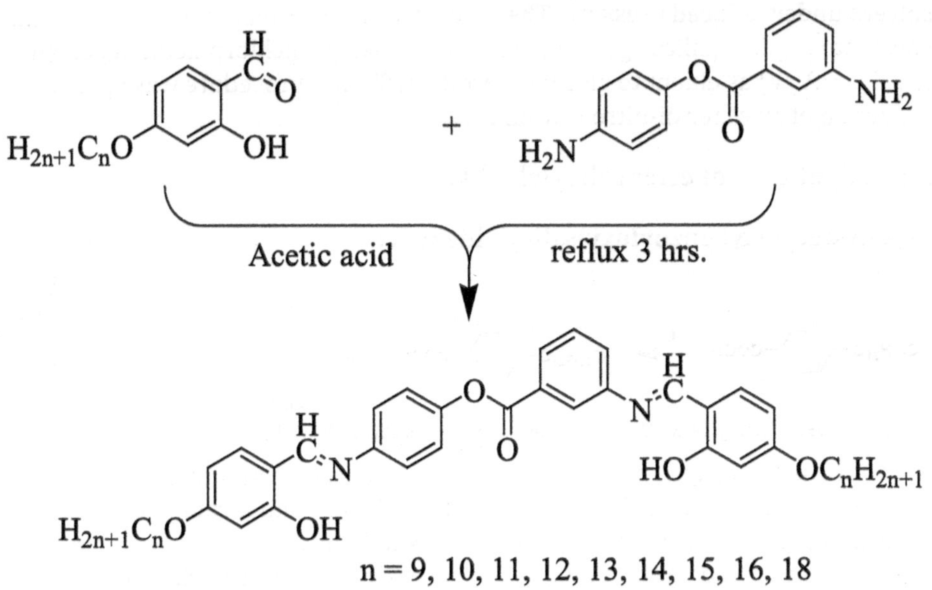

Scheme 16: Synthesis of RN-n

[4-(N-4-nonyloxysalicylidenemino)phenyl]-[3-(N-4-nonyloxysalicylideneimino) benzoate]: RN-9

An ethanolic solution of 4-aminophenyl-3-aminobenzoate (0.58 g, 2.5 mmol) was added to an ethanolic solution (20 ml) of 4-n-decyloxysalicyldehyde (1.32 g, 5 mmol). The mixture was refluxed with a few drops of glacial acetic acid as catalyst for 3 hours to yield the Schiff's base [4-(N-4/-n-nonyloxysalicylidenemino) phenyl]-[3-(N-4/-n-nonyloxysalicylideneimino)benzoate], (RN-**9**). The precipitate was collected by filtration from the hot solution and recrystallized several times from absolute ethanol to give a pure compound.

Using the same procedure as described above (Scheme 16) the whole homologous series with varying number of carbon atoms (n) in the alkoxy chain (n = 10, 11, 12, 13, 14, 15, 16, 18) were synthesized.

3.6.2. [4-(N-4-alkyloxysalicylidenemino)phenyl]-[2-methyl-3-(N-4-alkyloxysalicylidenemino) benzoate]: (RN-n-2Me)

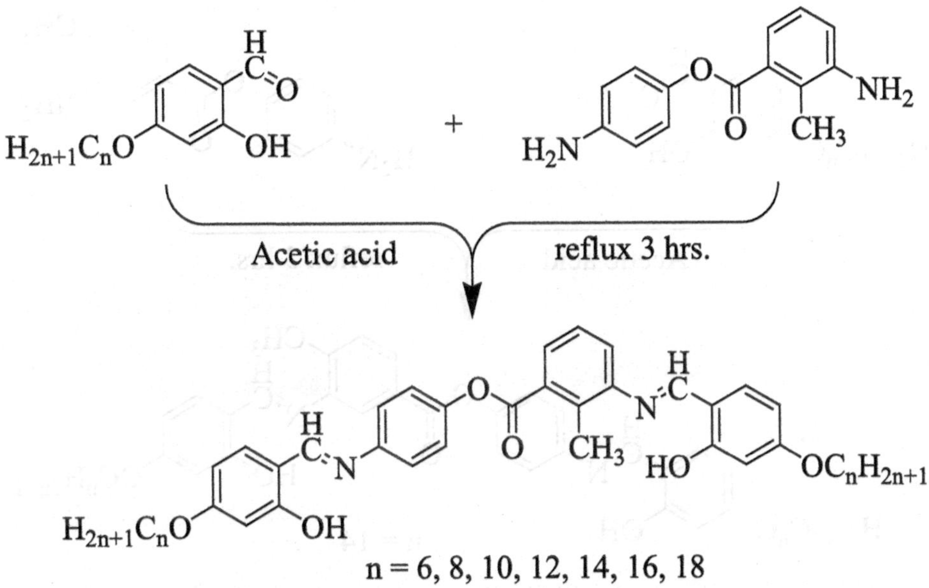

Scheme 17: Synthesis of RN-n-2Me

[4-(N-4-hexyloxysalicylidenemino)phenyl]-[2-methyl-3-(N-4-hexyloxysalicylidenemino) benzoate]: (RN-6-2Me)

An ethanolic solution of 4-aminophenyl-3-amino-2-methylbenzoate (0.6g, 2.5 mmol) was added to an ethanolic solution (20 ml) of 4-n-hexyloxysalicyldehyde (1.11 g, 5 mmol). The mixture was refluxed with a few drops of glacial acetic acid as catalyst for 3 hours to yield the yellow colored Schiff's base [4-(N-4/-n-hexyloxysalicylidenemino)phenyl]-[3-(N-4/-n-hexyloxysalicylidenemino) benzoate], (**RN-6-2Me**). The precipitate was collected by filtration from the hot solution and recrystallized several times from absolute ethanol to give a pure compound.

Using the same procedure as described above (Scheme 17) the whole homologous series with varying number of carbon atoms (n) in the alkoxy chain (n = 8, 10, 12, 14, 16, 18) were synthesized.

3.6.3. [4-(N-4-alkyloxysalicylidenemino)phenyl]-[4-methyl-3-(N-4-alkyloxysalicylidenemino) benzoate]: (RN-n-4Me)

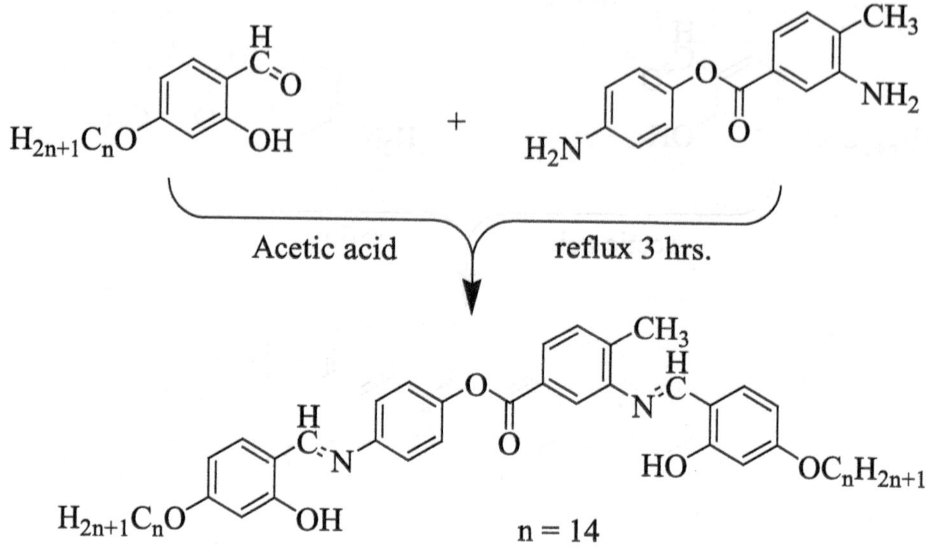

Scheme 18: Synthesis of RN-n-4Me

[4-(N-4-tetradecyloxysalicylidenemino)phenyl]-[4-methyl-3-(N-4-tetradecyloxysalicylidenemino)benzoate]: (RN-14-4Me)

An ethanolic solution of 4-aminophenyl-3-amino-4-methylbenzoate (0.6g, 2.5 mmol) was added to an ethanolic solution (20 ml) of 4-n-tetradecyloxysalicyldehyde (1.67 g, 5 mmol). The mixture was refluxed with a few drops of glacial acetic acid as catalyst for 3 hours to yield the Schiff's base [4-(N-4/-n-octyloxysalicylidenemino)-phenyl]-[4-methyl-3-(N-4/-n-octyloxysalicylidenemino) benzoate], **(RN-14-4/Me)**. The precipitate was collected by filtration from the hot solution and recrystallized several times from absolute ethanol to give a pure compound.

3.6.4. [4-(N-4-alkyloxysalicylidenemino)phenyl]-[4-chloro-3-(N-4-alkyloxysalicylidenemino)benzoate]: (RN-n-4Cl)

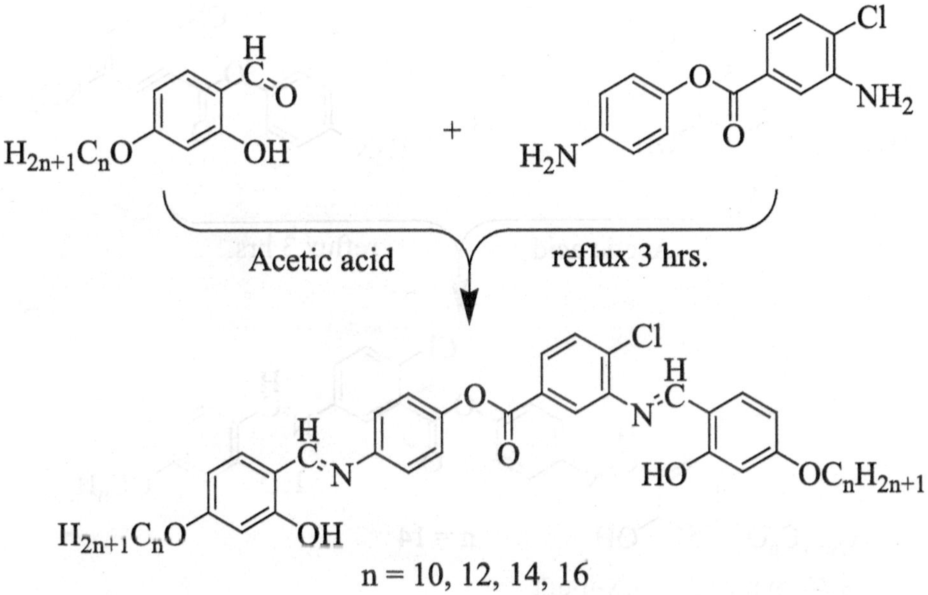

Scheme 19: Synthesis of RN-n-4Cl

[4-(N-4-decyloxysalicylidenemino)phenyl]-[4-chloro-3-(N-4-decyloxysalicylidenemino)benzoate]: (RN-10-4Cl)

An ethanolic solution of 4-aminophenyl-3-amino-4-chlorobenzoate (0.65 g, 2.5 mmol) was added drop wise to an ethanolic solution (20 ml) of 4-n-decyloxysalicyldehyde (1.39 g, 5 mmol). The mixture was refluxed with a few drops of glacial acetic acid as catalyst for 3 hours to yield the Schiff's base [4-(N-4/-n-decyloxysalicylidenemino)phenyl]-[4-chloro-3-(N-4/-n-decyloxysalicylidenemino)-benzoate], **(RN-10-4Cl)**. The precipitate was collected by filtration from the hot solution and recrystallized several times from absolute ethanol to give a pure compound.

Using the same procedure as described above (**Scheme 19**) the whole homologous series with varying number of carbon atoms (n) in the alkoxy chain (n = 12, 14, 16) were synthesized.

3.6.5. [4-(N-4-alkyloxysalicylidenemino)phenyl]-[2-chloro-5-(N-4-alkyloxysalicylidenemino) benzoate]: (RN-n-6Cl)

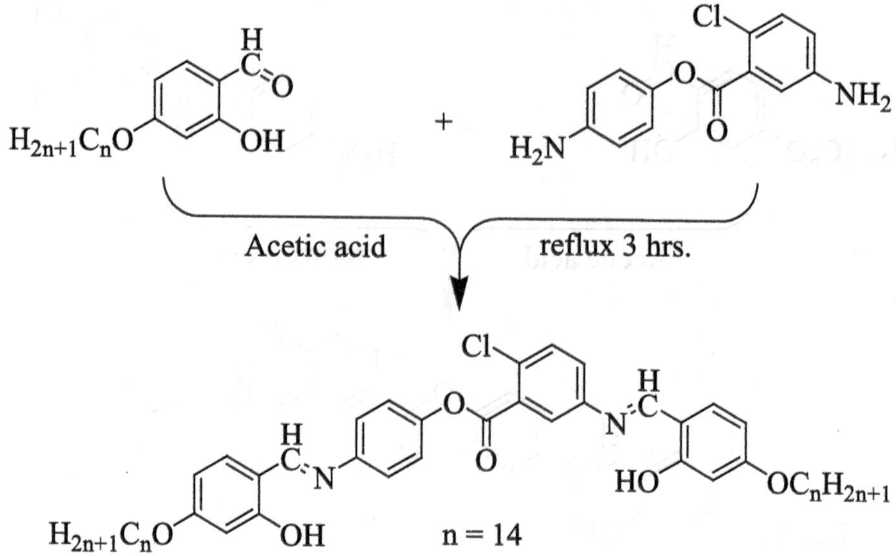

Scheme 20: Synthesis of RN-n-6Cl

[4-(N-4-tetradecyloxysalicylidenemino)phenyl]-[2-chloro-5-(N-4-tetradecyloxysalicylidenemino) benzoate]: (RN-14-6Cl)

An ethanolic solution of 4-aminophenyl-5-amino-2-chlorobenzoate (0.65 g, 2.5 mmol) was added drop wise to an ethanolic solution (20 ml) of 4-n-tetradecyloxysalicyldehyde (1.67 g, 5 mmol). The mixture was refluxed with a few drops of glacial acetic acid as catalyst for 3 hours to yield the yellow coloured Schiff's base [4-(N-4/-n-tetradecyloxysalicylidenemino)phenyl]-[2-chloro-5-(N-4/-n-tetradecyloxysalicylidenemino)benzoate], **(RN-14-6Cl)**. The precipitate was collected by filtration from the hot solution and recrystallized several times from absolute ethanol to give a pure compound.

3.6.6. [4-(N-4-alkyloxysalicylidenemino)phenyl]-[2-chloro-5-(N-4-alkyloxysalicylidenemino) benzoate]: (RN-10-6F)

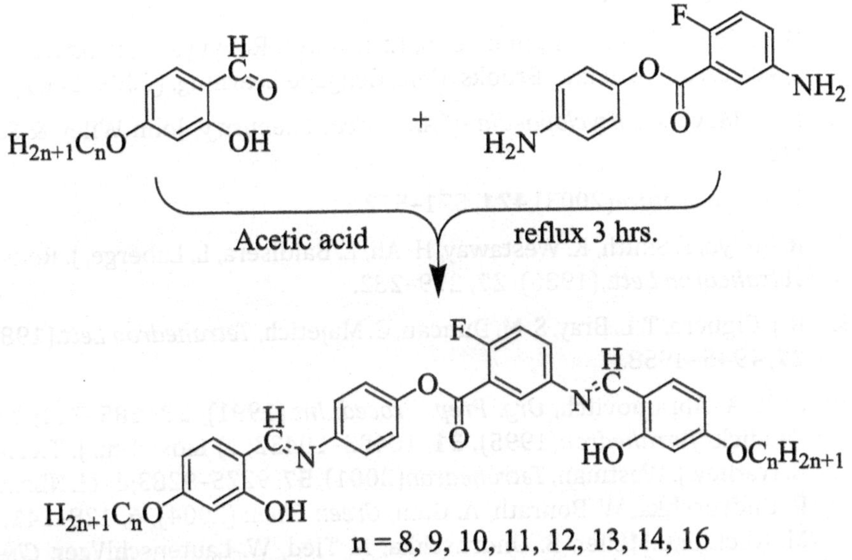

Scheme 21: Synthesis of RN-n-6F

[4-(N-4-decyloxysalicylidenemino)phenyl]-[2-chloro-5-(N-4-decyloxysalicylidenemino) benzoate]: (RN-10-6F)

An ethanolic solution of 4-aminophenyl-3-amino-6-fluorobenzoate (0.62 g, 2.5 mmol) was added drop wise to an ethanolic solution (20 ml) of 4-n-decyloxysalicyldehyde (1.39 g, 5 mmol). The mixture was refluxed with a few drops of glacial acetic acid as catalyst for 3 hours to yield the yellow coloured Schiff's base [4-(N-4/-n-decyloxysalicylidenemino)phenyl]-[2-fluoro-5-(N-4/-n-decyloxysalicylidenemino)benzoate], (**RN-10-6F**). The precipitate was collected by filtration from the hot solution and recrystallized several times from absolute ethanol to give a pure compound.

Using the same procedure as described above (**Scheme 21**) the whole homologous series with varying number of carbon atoms (n) in the alkoxy chain (n = 8,9,11,12,13 14,16) were synthesized.

References:

1. M. Ghedini, B. Panuzi and A. Roviello, *Liq. Cryst.* (1998), **25**, 225.
2. N. Hoshino, A. Kodama, T. Shibuya, Y. Matsunaga and S. Miyajima,*Inorg. Chem.* (1991), **30**, 3091.
3. S. F. Tandel, M.W. Schoder, G. Pelzl, U. Baumeister, W. Weissflog, S. Stern, A. Nemes, R. Stannarius, and A. Eremin, *Eur. Phys. J.* E, (2008), **25**, 395.

4. *Vogel's Textbook of Practical Organic Chemistry*, Ed B. S. Furniss, A. J. Hannaford, P. W. G. Smith and A. R. Tachell, Longman Scientific and Technical, UK 1989.

5. D. L. Pavia, G. M. Lampman, G. S. Kriz, and J. R. Vyvyan, *Introduction to Spectroscopy*, 4th Ed., Brooks/Cole, Cengage Learning, (2009, 2001)

6. R. A. Meyers, *Encyclopaedia of Analytical Chemistry*, John Wiley & Sons Ltd

7. D. Adam, *Nature*(2003),**421**, 571–572.

8. R. Gedye, F. Smith, K. Westaway, H. Ali, L. Baldisera, L. Laberge, J. Rousell, *Tetrahedron Lett.*,(1986), **27**, 279–282.

9. R. J. Giguere, T. L. Bray, S. M. Duncan, G. Majetich, *Tetrahedron Lett.*(1986), **27**, 4945–4958.

10. a) R. A. Abramovitch, *Org. Prep. Proced. Int.*(1991), **23**, 685–711; b) S. Caddick, *Tetrahedron*(1995), **51**, 10403–10432; c) LidstrUm, J. Tierney, B. Wathey, J. Westman, *Tetrahedron*(2001), **57**, 9225–9283;d) M. NLchter, B. Ondruschka, W. Bonrath, A. Gum, *Green Chem.*(2004), **6**, 128–141; e) M. NLchter,U.MLller, B. Ondruschka, A. Tied, W. LautenschlVger, *Chem. Eng. Technol.*(2003), **26**, 1207–1216.

11. a) C. R. Strauss, R.W. Trainor, *Aust. J. Chem.*(1995), **48**, 1665–1692; b) C. R. Strauss, *Aust. J. Chem.*(1999), **52**, 83–96.

12. a) A. K. Bose, B. K. Banik, N. Lavlinskaia, M. Jayaraman, M. S. Manhas, *Chemtech*(1997), **27**, 18–24; b) A. K. Bose, M. S. Manhas, S. N. Ganguly, A. H. Sharma, B. K. Banik, *Synthesis*(2002), 1578 – 1591.

13. A. de la Hoz, A. DWaz-Ortis, A. Moreno, F. Langa, *Eur. J. Org. Chem.*(2000), 3659 – 3673.

14. a) J. Hamelin, J.-P. Bazureau, F. Texier- *Boullet in Microwaves in Organic Synthesis* (Ed.: A. Loupy), Wiley-VCH, Weinheim, 2002, pp. 253 – 294; b) T. Besson, C. T. Brain in *Microwave-Assisted Organic Synthesis* (Eds.: P. LidstrUm, J. P. Tierney), Blackwell, Oxford, 2004, Chap. 3; c) Y. Xu, Q.-X. Guo, *Heterocycles*(2004), **63**, 903–974.

15. a) N. Elander, J. R. Jones, S.-Y. Lu, S. Stone- Elander, *Chem. Soc. Rev.* (2000), **29**, 239–250; b) S. Stone- Elander, N. Elander, J. Labelled *Compd. Radiopharm.*(2002), **45**, 715–746.

16. a) M. Larhed, C. Moberg, A. Hallberg, *Acc. Chem. Res.*(2002), **35**, 717–727;

17. a) J. L. Krstenansky, I. Cotterill, *Curr. Opin. Drug Discovery Dev.*(2000), **3**, 454–461; b) M. Larhed, A. Hallberg, *Drug Discovery Today*(2001), **6**, 406–416; c) B. Wathey, J. Tierney, P. LidstrUm, J. Westman, *Drug Discovery Today* (2002), **7**, 373–380; d) N. S. Wilson, G. P. Roth, *Curr. Opin. Drug Discovery Dev.*(2002), **5**, 620–629; e) C. D. Dzierba, A. P. Combs, *Ann. Rep.*

Med. Chem., Vol. 37 (Ed.: A. M. Doherty), Academic Press, 2002, pp. 247 – 256.

18. a) A. Lew, P. O. Krutznik, M. E. Hart, A. R. Chamberlin, *J. Comb. Chem.* (2002), **4**, 95–105; b) C. O. Kappe, *Curr. Opin. Chem. Biol.* (2002), **6**, 314–320; c) P. LidstrUm, J. Westman, A. Lewis, *Comb. Chem. High Throughput Screening* (2002), **5**, 441–458; d) H. E. Blackwell, *Org. Biomol. Chem.* (2003), **1**, 1251–1255; e) F. Al-Obeidi, R. E. Austin, J. F. Okonya, D. R. S. Bond, *Mini-Rev. Med. Chem.* (2003), **3**, 449–460; f) K. M. K. Swamy, W.-B. Yeh, M.-J. Lin, C. M. Sun, *Curr. Med. Chem.* (2003), **10**, 2403–2423; g) *Microwaves in Combinatorial and High-Throughput Synthesis* (Ed.: C. O. Kappe), Kluwer, Dordrecht, 2003.
19. *www.maos.net*
20. *Microwaves in Organic Synthesis* (Ed.: A. Loupy),Wiley-VCH, Weinheim, 2002.
21. B. L. Hayes, *Microwave Synthesis: Chemistry at the Speed of Light*, CEM Publishing, Matthews NC, 2002.
22. *Microwave-Assisted Organic Synthesis* (Eds.: P. LidstrUm, J. P. Tierney), Blackwell, Oxford, 2004.
23. D. R. Baghurst, D. M. P. Mingos, *Chem. Soc. Rev.* (1991), **20**, 1–47.
24. C. Gabriel, S. Gabriel, E. H. Grant, B. S. Halstead, D. M. P. Mingos, *Chem. Soc. Rev.*(1998), **27**, 213–223.

CHAPTER 4

Unsubstituted Four-ring Bent Core Mesogen

This chapter describes the detailed studies of the unsubstituted four ring bent-core liquid crystals. Their phase behaviour, electro-optic and emissive properties are described. Compounds of the present study show polarisation modulated layer undulated (PMLU) phase structure, which are discussed elaborately.

4. Unsymmetrical 4-ring bent core mesogens:

The four-ring system with a short odd aliphatic chain spacer is a dimesogen and exhibits polar/chiral phases based on the bent shape promoted by the spacer. [1] However the molecules with only four rings in the core with a biphenyl moiety of 1,5-disubstituted-2,3,4-trifluorophenyl biphenylcarboxylates [(S)-1-methylheptyl 4-[2,3,4-trifluoro-5-(4/-dodecyloxy-biphenyl-4-ylcarbonyloxy)-phenylcarbonyloxy] benzoate] and exhibited conventional SmA, ferro-, ferri-, and antiferroelectric SmC phases resulting from the molecular chirality and not due to the bent shape[2]. The modification of angular 3,4/-disubstituted biphenyl unit with the introduction of an ester linkage viz., COO moiety between the phenyl units leads to structural variation of polar groups in bent core molecules. The ester linkage as well as reduction of phenyl rings in general reduces the melting and clearing temperatures. It had been demonstrated between the five and four ring systems earlier. *How the reversal of the central ester linkage can influence the phase structure and phase transition temperatures?* To examine the effect of direction of the ester linkage on the phase transitions and phase structure, a homologous series of compounds are synthesized and their characteristics are studied. The molecular structure of the homologous series is presented below. Moreover, with respect to the unequal distribution of the aromatic rings in the two wings these molecules could also be regarded as hockey stick molecules. The salicylideneimine unit has been previously reported as a useful structural unit for bent-core mesogens [3-6]. Its use leads to an extended conjugation in the core and, thereby, an additional attractive property of the new cores reported here is their strong photoluminescence, opening the way to use their fluorescence characteristics in the structural investigation of bent-core mesomorphic phases.

4.1. Compounds designed and synthesized:

The bent core unit viz. ***3,4′-disubstituted biphenyl*** (**4-1**)had been successfully demonstrated as a central core unit in five, six and seven ring molecules exhibiting several variants of banana phases. The modification of 3,4′-disubstituted biphenyl central unit with the introduction of an ester linkage between the phenyl rings can yield two different central cores 3-aminophenyl-4-aminobenzoate (**4-2**) and 4-aminophenyl-3-aminobenzoate (**4-3**) which can be subsequently coupled with 4-n-alkoxysalicyliladehydes to obtain the designed compounds. In this section the results on the homologous series of [4-(N-4′-n-alkyloxysalicylidene)-aminophenyl]-[3-(N-4′-n-alkyloxysalicylideneamino)-benzoate with varying the number of carbon atoms in alkyl moiety from 9 to 16 and 18 (**RN-n,** with **n = 9-16,** and **18**) are presented.

Modified Structures of central bent core unit

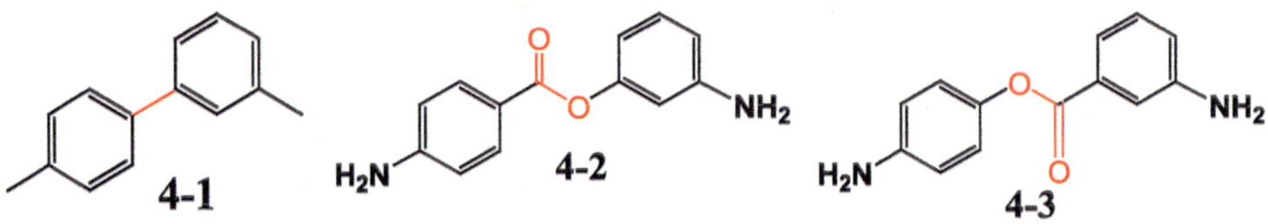

4-1 — 3,4′-disubstituted biphenyl

4-2 — 3-aminophenyl-4-aminobenzoate

4-3 — 4-aminophenyl-3-aminobenzoate

Designed compound

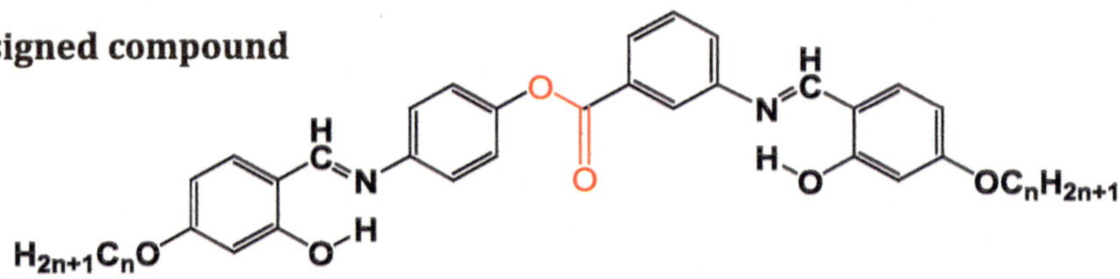

RN-n: **n = 9, 10, 11, 12, 13, 14, 15, 16, 18**

[4-(N-4′-n-alkyloxysalicylidene)-aminophenyl]-[3-(N-4′-n-alkyloxysalicylideneamino)-benzoate

Figure 4.1: Design of the molecule

4.2.[4-(N-4′-n-decyloxysalicylidene)aminophenyl]-[3-(N-4′-n-decyloxysalicyli- dene amino)-benzoate (RN-10):

4.2.1. Synthesis:

The crucial prerequisite for achieving polar order and chirality in any of the materials is that the individual constituent molecule should possess a bent shape in the central core and non-coplanarity of the phenyl rings in the molecule to result into noncentrosymmetric architecture. In attempting to devise a strategy for the synthesis of bent-shaped molecules, we reasoned that Carsten's angular 3,4′-disubstituted biphenyl central unit [7, 8] can be separated by an ester linkage which can still be sufficiently bent shaped and unsymmetrical to match the perturbation produced in the form of a shape factor by the introduction of a bridging group between two phenyl rings to yield a unsymmetrical molecule. The tail end of this unusual bent-core liquid crystal was based on 4-n-alkoxysalicylaldehyde moieties where as bent core platform was designed by the coupling of 4-nitrophenol with 3-nitrobenzoic acid. The starting material in the present study 4-n-alkyloxysalicylaldehyde was prepared by Williamson etherification of 2,4-dihydroxybenzaldehyde with appropriate n-alkyl bromide. The central bent core moiety 3-aminophenyl-4-aminobenzoate was prepared as follows. 3-nitrobenzoic acid was converted into its acyl chloride followed by condensation with 4-nitrophenol using phase transfer reaction. The resultant 4′-nitrophenyl-3-nitrobenzoate was subjected to 10% Pd-C catalysed reduction to get the desired di-amino bent-core compound. The condensation of 4-n-alkyloxy salicylaldehyde with 4-aminophenyl-3-aminobenzoate in presence of a few drops of glacial acetic acid yielded the target bent shaped compounds [4-(N-4′-n-decyloxysalicylidene)-aminophenyl]-[3-(N-4′-n-decyloxysalicylideneamino)-benzoate (**RN-n, n = 10**). To avoid the formation of side products, the precipitated compounds were filtered when the solution was hot to yield the pure compounds. The compounds were further recrystallized repeatedly to get the pure samples. [9] The formation of all of the compounds was confirmed by ^{1}HNMR and IR spectroscopy and the purity was established by elemental analysis. The other homologues with n = 9, 11, 12, 13, 14, 15, 16 and 18 were also synthesized following the same procedure and characterized. The details of the experimental procedures along with the spectroscopic data for all the homologous series of compounds are presented in the experimental part of Chapter 3. The liquid-crystalline behaviour of the synthesised compounds had been investigated by optical microscopy and differential scanning calorimetry (DSC).

4.2.2. Mesomorphic properties:

The transition temperatures, enthalpies and entropies associated with the phase transitions of all the compounds of the homologous series **RN-n** (n = 9 to 16, 18) as a function of number of the carbon atoms in the terminal alkyl chains obtained from DSC at a scan rate of 5^{0}C min^{-1} in the second heating and first cooling scans are presented in **Table 4.1.** All compounds of the homologous series **RN-n**exhibitbananaliquid crystalline behaviour over a moderate temperature region. The calorimetric study on a representative compound RN-10 is presented here. **RN-9** exhibited only monotropic liquid crystalline phase in the cooling cycle. **RN-10** exhibited two phase transitions in heating and cooling cycles.

Crystal 135.38^{0}C (ΔH = 47.87 kJ/mol, ΔS = 117.22 J/K/mol) LC 139.59^{0}C (ΔH = 11.45 kJ/mol, ΔS = 27.75 J/K/mol) Isotropic phase. In the cooling cycle also it exhibited enantiotropic phase transitions at 137.81^{0}C (ΔH = 14.18 kJ/mol, ΔS =

34.51 J/K/mol) and 111.97°C (ΔH = 43.03 kJ/mol, ΔS = 111.77 J/K/mol). Three crystal to crystal phase transitions at 83.34°C, 90.86°C and 97.38°C in the first heating cycle and at 79.4°C in the cooling cycle are also observed which had not been shown. The differential scanning calorimetric thermogram of **RN-10** is presented in **Figure 4.2**.

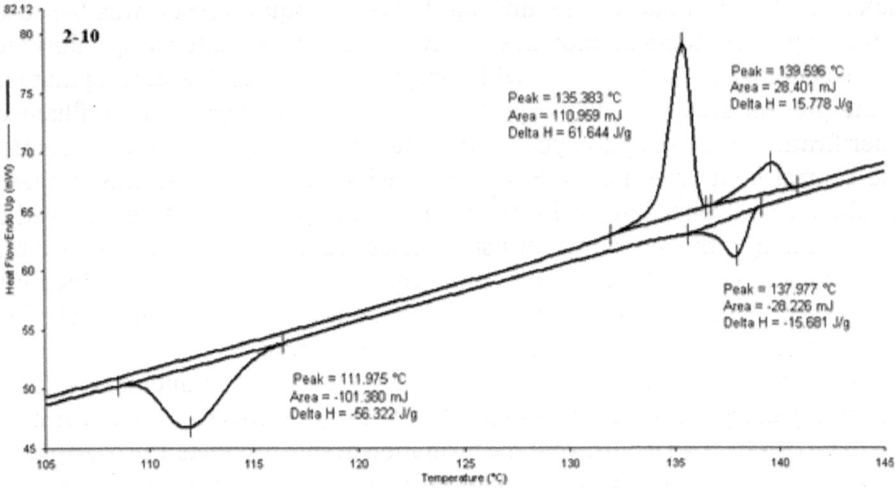

Figure 4.2: DSC thermogram of RN-10

The phase transition temperatures as a function of number of carbons in the end alkyl chain length is depicted in **Figure 4.3**. All the compounds were found to exhibit one mesophase only exhibiting similar textures. The compound with carbon atoms n = 14 and 10 in the alkyl chain length had exhibited wide mesophase thermal range whereas compound with carbon atoms n = 9 in the alkyl chain length exhibited smallest monotropic mesophase thermal range (**figure 4.4**).

Table 4.1: Transition Temperatures (°C), Enthalpies (*Italic*,kJ/mol) and entropies (*Italic*, J/mol/K) [shown in parantheses] of homologous series **RN-n**.

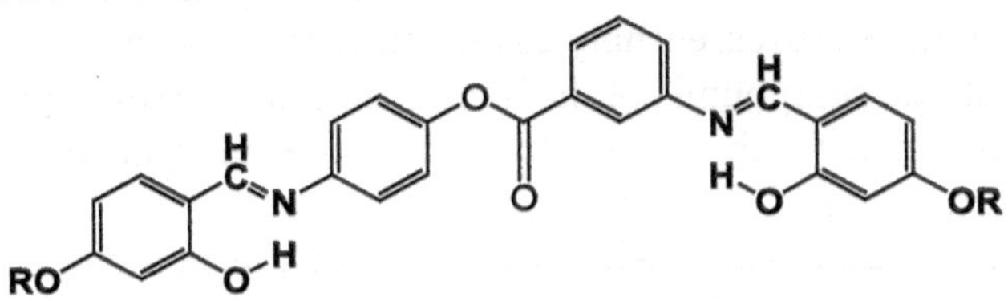

Compound	R	Cr	Heating Cooling	B1	Heating Cooling	I	ΔT
RN-9	C_9H_{19}	•	-	•	138.52 [*54.4, 132.2*]	•	
		•	124.15 [*41.8, 105.4*]	•	131.23 [*4.5, 11.1*]	•	7.1

RN-10	$C_{10}H_{21}$	•	135.38 *[47.9, 117.2]*	•	139.59 *[11.4, 27.7]*	•	4.2
		•	111.97 *[43.0, 111.8]*	•	137.8 [14.2, *34.5*]	•	25.9
RN-11	$C_{11}H_{23}$	•	136.53 *[55.8, 136.2]*	•	140.39 *[15.0, 36.3]*	•	3.8
		•	118.25 *[52.8, 135.1]*	•	139.20 *[19.2, 46.6]*	•	21.0
RN-12	$C_{12}H_{25}$	•	136.26 *[49.8, 121.7]*	•	141.12 *[15.4, 37.2]*	•	4.9
		•	116.71 *[43.1, 110.6]*	•	139.41 *[15.4, 37.3]*	•	22.7
RN-13	$C_{13}H_{27}$	•	133.54 *[36.6, 90.1]*	•	139.33 [11.4, *27.6*]	•	5.8
		•	114.86 *[34.4, 88.8]*	•	138.30 *[10.0, 24.3]*	•	23.5
RN-14	$C_{14}H_{29}$	•	133.02 *[44.0, 108.3]*	•	139.89 *[12.3, 29.7]*	•	6.8
		•	111.64 *[40.7, 105.8]*	•	138.61 *[15.4, 37.4]*	•	27.0
RN-15	$C_{15}H_{31}$	•	131.17 *[48.6, 120.3]*	•	138.54 *[17.1, 41.7]*	•	7.4
		•	114.74 *[48.2, 124.4]*	•	137.51 *[16.8, 41.0]*	•	22.8
RN-16	$C_{16}H_{33}$	•	132.68 *[64.1, 157.9]*	•	137.80 *[20.1, 48.8]*	•	5.2
		•	114.91 *[56.4, 145.3]*	•	136.53 *[18.7, 45.7]*	•	21.6
RN-18	$C_{18}H_{37}$	•	130.05 *[56.0, 12.9]*	•	134.34 *[15.9, 39.1]*	•	4.3
		•	111.22 *[86.1, 224.0]*	•	133.15 *[15.8, 39.0]*	•	21.9

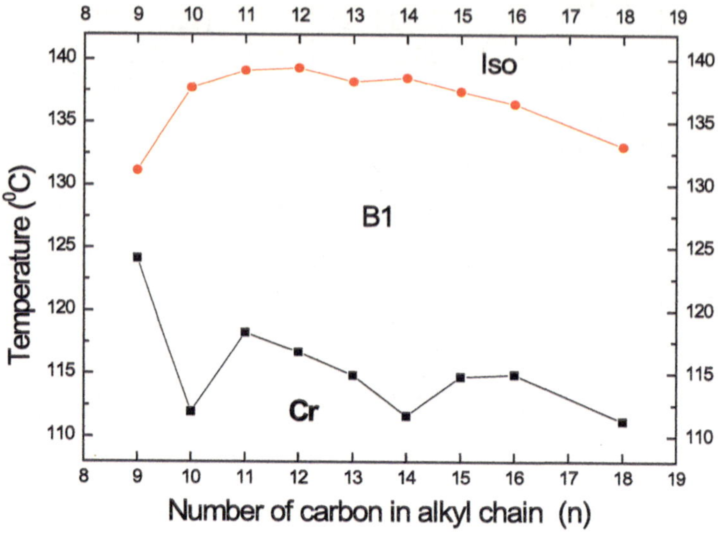

Figure.4.3: Phase transition temperatures as a function of number of carbons in the alkyl chain length in the cooling cycle

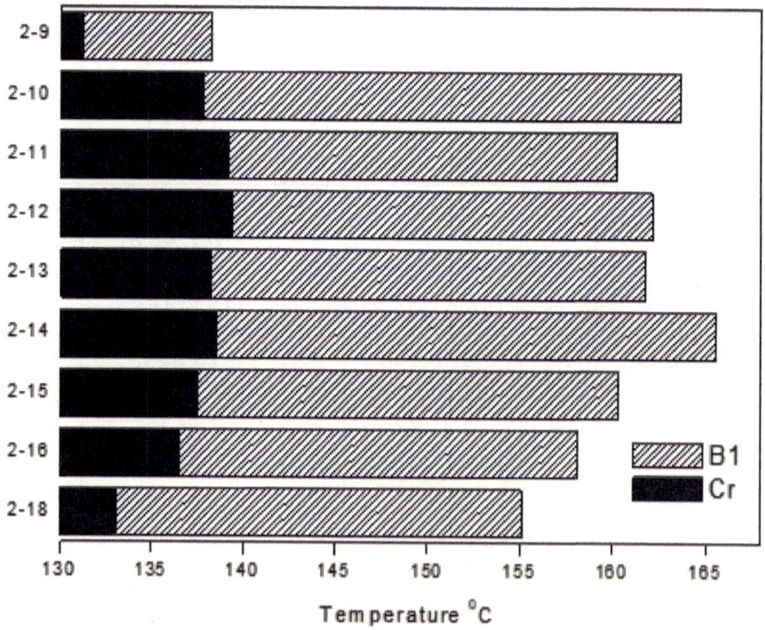

Figure.4.4: Mesomorphic range as a function of number of carbons in the alkyl chain length

The observed optical textures exhibited by a representative compound[4-(N-4/-n-decyloxysalicylidene)-aminophenyl]-[3-(N-4/-n-decyloxysalicylideneamino) benzoate**RN-10** are presented in **Figure 4.5.** Upon very slow cooling the isotropic liquid, compounds RN-n with n = 10 exhibited developable domains (**Figure 4.5a**) and telephone-wire like spiral filaments (**Figure 4.5b** and **c**), circular domains, and banana leaf-like domains. The developable domain-like textures resemble the rectangular columnar B1 phases. The growth of telephone-wire like spiral filament textures has been widely related to the existence of smectic undulated phases.[10-12] However recent investigations based on high resolution X-ray studies revealed that such type of textures are exhibited by 2D modulated smectic phases of bent core molecules based on locally ferroelectric layering and spontaneous splay of the polarization. Based on polarization splay modulation (either syn- or anti-order), slopes of undulated layers and layer continuity versus layer discontinuity at the intervening defects twelve different phases are proposed. Only some of the phase structures had been realized till now. The textural features of the compounds in the present work **(RN-n)** resemble the textures observed in compounds exhibiting 2D polarization splay modulated layer undulated B7/$B1_{RevTilted}$ phases [10-12].

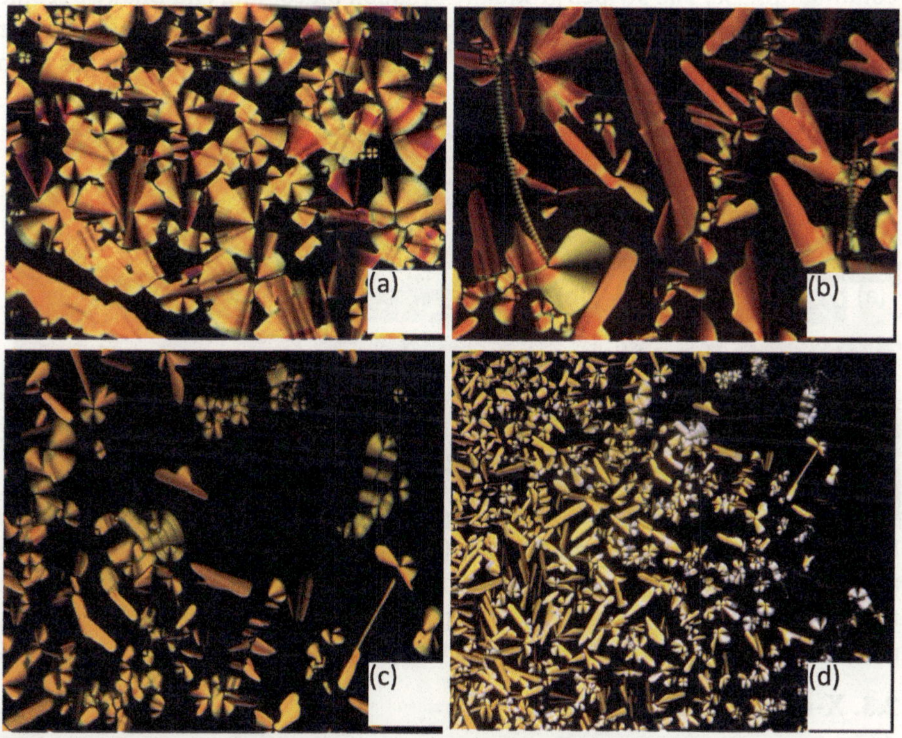

Figure 4.5a: Optical microphotographs of **RN-14** at T = 135°C under crossed polarizers in different regions (a) developable domains (b) telephone wire like spiral texture with banana leaflike (c) circular domains and spiral texture (d) developable domains as well as spiral textures.

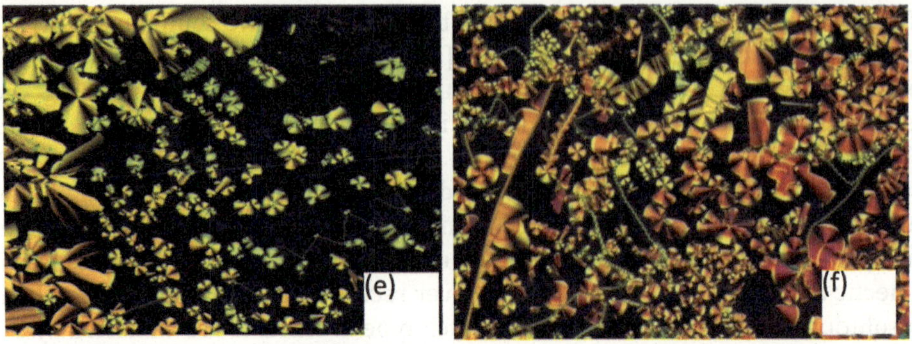

Figure 4.5b: Optical microphotographs of **RN-10** under crossed polarizers (e) circular domains and spiral texture at 134°C (f) telephone wire like spiral texture with banana leaflike at 131°C.

Optical photographs of **RN-14** were taken on homogeneous and homeotropically aligned glass surface. In homogenous cell (HTAB treated cell) of the compound **RN-14**, blue fan shaped defects of characteristic B1 phase appear on cooling from isotropic liquid. On further cooling, violet and pink stripes appears on the blades of columnar phase. The stripes gradually broaden on cooling and finally become violet/pink before crystallisation. In planer cell (nylon 6,6 treated cell) circular fan appears along with columnar fan. On cooling, colour of the fan changes from green to pink **(Figure 4.6)**.

Figure 4.6: Optical photomicrograph (crossed polarizers) of the B1 phase as obtained by cooling from the isotropic liquid of **RN-14**. (a), (b), (c) HTAB treated cell at 132.5, 118 and 117 °C respectively. (d), (e), (f) Nylon 6,6 treated cell at 133.4, 119.5 and 117.8 °C respectively.

4.2.3. X-ray studies:

RN-10

The powder X-ray diffraction patterns of the liquid crystalline phase of the compound **RN-10** is displayed in **Figure 4.7a**. In the small angle region compound **RN-10** exhibited one sharp layer reflection with a spacing of 49.0 Å (T = 131.8°C) which is equal to the length of the molecule in the most extended conformation (L = 48.9 Å, **Figure 4.8a** and **b**) inferring that the molecules are arranged in orthogonal monolayer structure. The outer broad reflections in the wide angle region **Figure 4.7b,** with multiple peaks at 4.33 and 4.97Å reflect the mean distance between the aliphatic chains with in each layer and suggest partial ordering in lateral packing of the molecules as seen in smectic phases. The temperature variation of layer thickness of compound **RN-10,** which is found to be constant, is shown in **Figure 4.6c.**

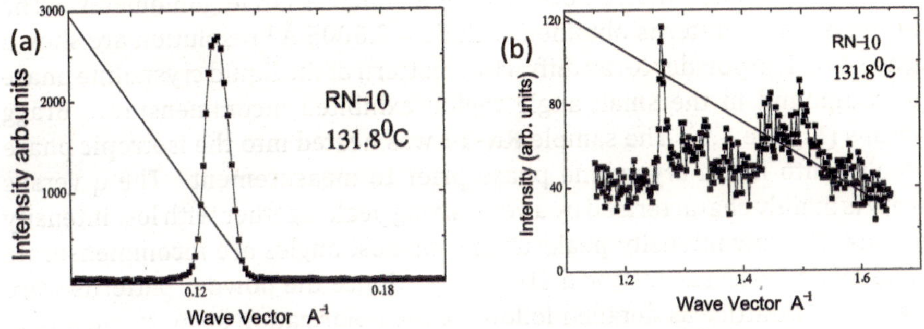

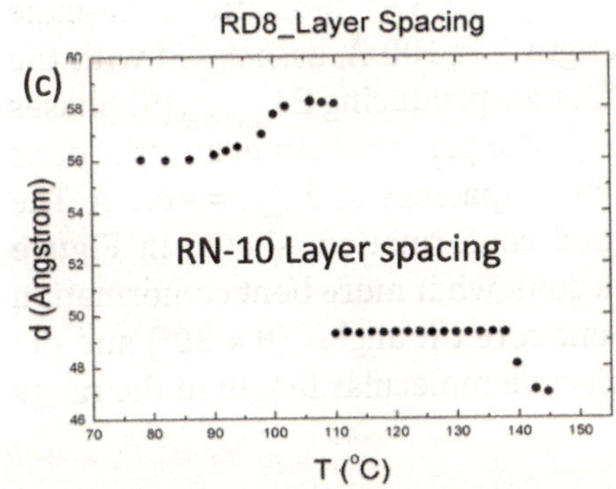

Figure 4.7: Powder XRD pattern of compound **RN-10** at 131.8°C, indicating a $B1_{RevTilted}$ phase structure (a, b). Figure (c) shows the variation of layer thickness as a function of temperature.

Fully extended all trans conformation and energy minimized space filling model of RN-10 is shown in **figure 4.8**. The length of the molecule is found to be 4.89nm (theoretical)

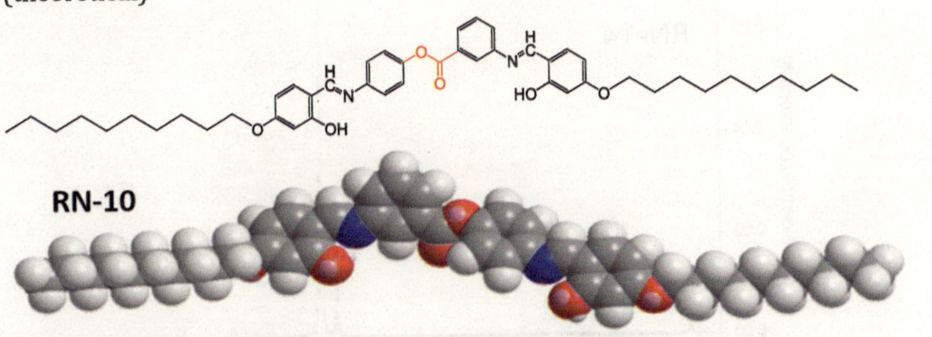

Layer spacing = 4.9nm Theoretical = 4.89nm

Figure: 4.8: a)Molecular structure and b) Space filling model of energy minimized structure of the molecule with fully extended end chains.

RN-14

To understand the phase characteristics in these compounds synchrotron XRD studies were carried out on powder sample of **RN-14** homologue of **RN-*n*** homologous series. The detailed X-ray scattering studies were performed on unaligned powder samples of **RN-14** contained in a 0.7 – 1.0 mm diameter 10 μm thick glass capillary at the X10A beam-line of National Synchrotron Light Source at Brookhaven National laboratory, USA. Samples were mounted in a temperature controlled Instec hot-stage on a Huber 4-circle goniometer. The resolution of X-ray patterns obtained with $\delta q = 0.0005$ Å^{-1} resolution are shown in **Figure 4.9**. The powder X-ray diffraction pattern of the liquid crystalline phase of the compound in the small angle region exhibited incommensurate Bragg reflections (**Figure 4.9**). The sample **RN-14** was heated into the isotropic phase and cooled into liquid crystalline phase prior to measurement. The q versus Intensity is mainly characterised by a very strong peak together with low intensity reflections. The low intensity peaks at the smallest angles are incommensurate with the other peaks suggesting a 2D lattice. Hence the powder patterns were indexed to a 2D lattice as detailed following the polarization splay stripe model [13]. The material shows a small-q reflection at $q \sim 0.06$ Å^{-1}, indicative of in-plane modulation of the smectic layers of wavelength $\lambda \sim 100$ Å, associated with the polarization splay modulation, a structural feature producing $B1_{RevTilted}$ [5] phases with a 2D lattice of ≈ 100 Å^{-1} (along the layer). The principal lamellar reflections at $q_{2\text{-}14} = 0.136$ Å^{-1} for RN-14 give smectic layer spacings of $d_{RN-14} = 46.1$ Å. The length of the molecule in the most extended conformation, shown in **Figure 4.9(c)**, is L ~ 59 Å, although it is likely that a somewhat more bent conformation is adopted in the smectic phases. Typical bent core tilt angles ($\theta \approx 30^0$) and the measured layer spacings would imply an effective molecular length in the range 50 Å < L < 54 Å.

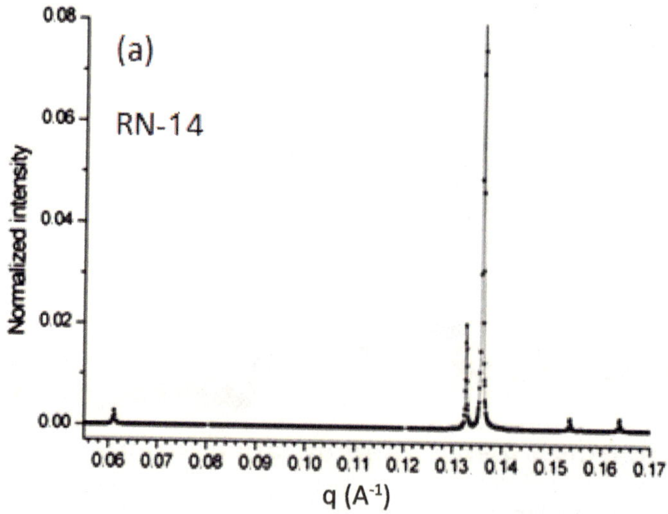

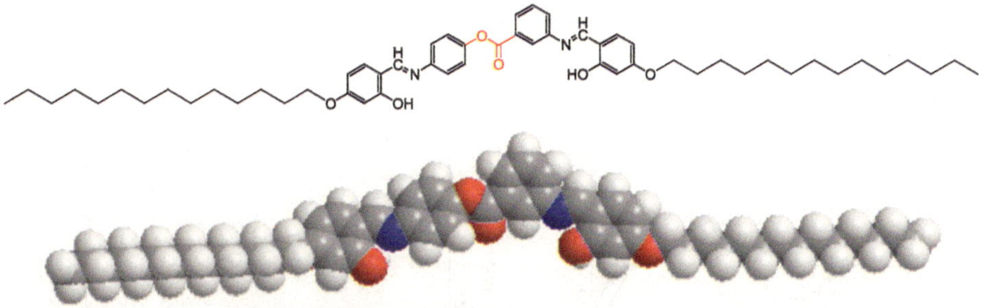

molecular length = 58.4 Å (theoretical)

Figure 4.9: (a) Compound **RN-14** powder XRD pattern, indicating a $B1_{RevTilted}$ phase structure. The material shows a small angle reflection at ~ 0.06 $Å^{-1}$, indicative of layer modulation of wavelength 1–100 $Å^{-1}$ associated with polarization splay. (b, c) fully extended all trans conformation and space filling model of minimum energy structure resp.

The X-ray diffraction studies with a strong Bragg reflection augmented by a number multiple peaks with in-plane modulation wave length of ~100 Å, twisted helical filament and striped domain POM textures suggested that the present compounds of the homologous series can be classified as SmCPU phases (which will be described in the discussion). These reflections cannot be assigned in absence of data on perfectly oriented samples but indicates a two-dimensional structure, complicated by a helicoidal superstructure as shown by the spiral defects observed in microscopic textures. Further work is in progress about understanding the molecular arrangement in this complex phase structure and to identify the sub class of the phase structure.

4.2.4. Electro-optical investigations:

To investigate the polar properties of the mesophase, electric field experiments were carried out on representative compound **RN-10**. The switching behaviour was investigated in transparent sandwich type capacitor cells (5 μm) consisting of two indium-tin oxide coated glass plates using the triangular wave method. Compound **RN-10**exhibitedone broad polarisation peak in the triangular wave even at a very low frequency of only 1 Hz. which is a strong indication of FE switching (P_S = 218 nC/cm^2). The current peak disappeared in isotropic phase. The mesogen exhibit ferroelectric smectic phase, characterized by the appearance of single polarization current peak per half period of the applied triangular wave voltage, with spontaneous polarization values ranging from 135 to 218 nC/cm^2 for compound **RN-10** as shown in **Figure 4.10a.**The value of spontaneous electric polarization is comparable to other banana phases. With decrease in temperature, the value of spontaneous polarization increases (**Figure 4.10b**). There is no significant change in texture (**Figure 4.11**) at the voltage at which the switching current occurred. Upon increasing the dc electric field a continuous increase in birefringence was observed reflecting the increase in orientational order of the molecules.

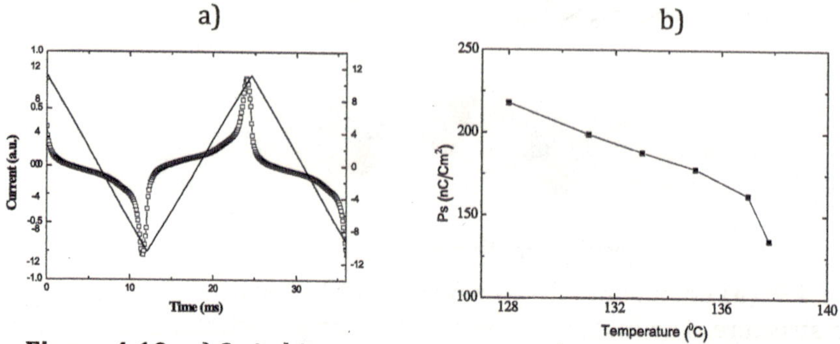

Figure 4.10: a) Switching current response for compound **RN-14.** b) Spontaneous polarization variation with temperature.

0V 30V 40V

Figure 4.11: Typical texture of **RN-14** in a 5μm cell at 135°C in B1 phase. The colour of the fan-shaped domain texture changes from blue to green on application of DC electric field.

4.2.5. Emissive Liquid Crystals, Absorption and Emission characteristics:

The incorporation of emission properties into organic liquid crystals (LCs) is currently an active area of research, which shows great promise. Emissive liquid crystals (LCs) are intriguing systems for materials research because they couple molecular self-assembly with intrinsic light generation capabilities. LCs offers excellent control over important parameters such as orientation, organization, geometry, and symmetry on the molecular level [14]. In addition, this control on the microscopic level can be extended into the macroscopic regime through appropriate processing and bulk alignment as in the case of LC displays. The effects of molecular anisotropy and organization on emission properties are not only an area of fundamental interest but also a means of constructing new functional materials with tailored anisotropic optical properties. Early work in the area of emissive LCs involved the incorporation of fluorescent dyes into conventional LC phases of rod-like molecules such as the nematic and smectic thermotropic phases. The orientation of fluorescent dyes in LC media has been used to generate functional organic materials with new anisotropic properties such as polarized organic lasers [15] and coloured, "active" guest-host LC displays [16-19]. A number of interesting phenomena have been observed in fluorescent dye-LC systems. For example, changes in the absorption and emission profiles of dyes dissolved in a nematic LC has been observed upon application of an electric field [20, 21]. Also, the fluorescence behaviour of oriented dyes in LC mixtures has been used to probe the nature of the ordered host media [22-24]. However, one of the common problems encountered with organic dye-LC mixtures is limited solubility of the dye in the LC host, especially in the case of fluorescent dyes [18-20]. The reported LC materials with intrinsic luminescence properties are meagre. Rod-like nematic mesogens based on fluorescent cyanobiphenyls [25, 26], benzopyran-4[H] [27] and oligo(p-phenylenevinylene) [28] cores have been investigated. Apart from that, columnar phases of discotic LCs containing polyaromatic cores have been synthesized and investigated with regard to fluorescence [29], unidimensional photoconductivity [30], and anisotropic electroluminescence for display applications [31, 32]. Phasmidic or polycatenar LCs with intrinsic luminescence properties were also studied intensively in respect of their organization in LC phase and its effect on emission profile.

The UV absorption and fluorescence spectroscopic properties of compound **RN-14** in solution were studied in various solvents of different concentrations to obtain information regarding absorption and emission maxima, and the Stokes shift of fluorescence. The compounds exhibited similar absorptions with absorption peaks at 344 nm (3.60 eV for RN-14, $\varepsilon \sim 50000$ $Lmol^{-1}cm^{-1}$). These absorption band with large molar absorption coefficients reflects the π-π^* transition of the highly π-conjugated system having the substituted phenyl benzoate unit as the core. Additionally, we investigated the fluorescence spectrum to observe the characteristic features of the excited states of **RN-14** .

The fluorescence emission spectra of these compounds were examined in chloroform solution (conc. 4.66 x 10^{-6} M) and it was found that both exhibited strong fluorescence on excitation at 350 nm. The emission peak at λ_{em} = 506 nm (2.45 eV) for compound RN-14, with a large Stokes shift λ_s of the order of λ_s = 162 nm (1.15 eV) was attributed to the formation of intermolecular excimers. This Stokes shift, which reflects the structural relaxation of the excited molecule, is significantly larger than in reported push-pull systems exhibiting liquid crystal behaviour,[33] indicating that the molecular conformation changes upon excitation. The plots of the fluorescence spectra at different concentrations show that the relative intensity of the peak at 515 nm gradually diminishes with dilution, indicating the formation of an intermolecular excimer (**Figure 4.12**). These results are in good agreement with the reported results of molecular J-aggregates [34] in which the excitonic energy is delocalized as a result of intermolecular coupling within the head-to-tail arrangement of the molecules in the solution. Both the compounds also exhibit fluorescence in the solid state. The optical band gaps E_g of both the compounds were determined from their absorption in thin solid films following the method in reference [35]. The E_g value obtained from the UV spectra is 3.35eV for **RN-14**. Fluorescence confocal polarizing microscope studies in these compounds confirmed the polarization dependent fluorescence in the mesophase.

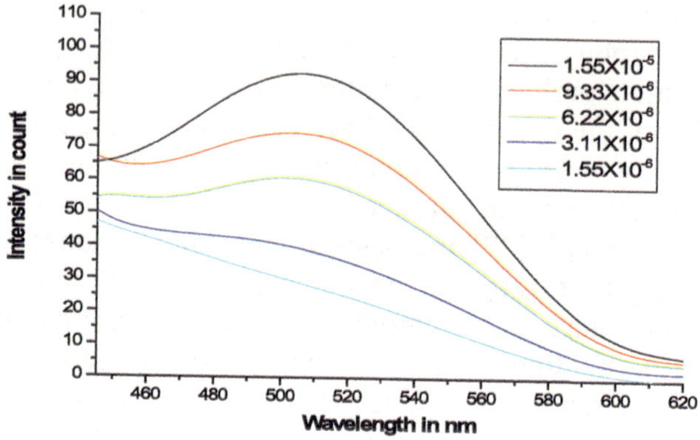

Figure 4.12: Concentration dependent Florescence spectra of **RN-14** on excitation at 365 nm in $CHCl_3$

4.3. Discussion:

The columnar mesophases are generated from two types of molecules viz. disc like molecules [36] and rod like molecules in which circular columns are formed by smectic layer fragments [37-40]. The bent core molecules are achiral but they form chiral superstructures. The bent core molecules because of bent shape exhibit polar order and some of them possess tilt order. In such polar LCs the polarization vector is free to rotate continuously in the layer plane under the influence of external fields. To escape from macroscopic polar order of the molecular structure the splay of polarization stabilizes the polarization modulation. The spontaneous breaking of chiral symmetry leads to a coupling between the polarization splay and molecular tilt order [41, 42]. Hence the coupling disturbs the homogeneous layered structure to manifest two-dimensional modulated structures viz. the polarization modulated either with a continuous periodically undulated smectic layers (PMLU, Classical B_7 phases) or with a discontinuous structure of blocks (columns) of smectic layer fragments ($B_{1RevTilt}$) organized into two-dimensional columnar like lattices. Recently it had been shown that banana molecules exhibiting rectangular columnar phases, manifested from broken layers, can possess spontaneously ordered dipoles [43-50]. The broken layers or the fragmented layers can form rectangular columns. The molecules in the layer fragments are organized with the polarization vector of adjacent fragments arranged in anti parallel direction. The shape of the banana molecules prevents the free rotation of molecules around its molecular long axis (a line joining the banana ends) and hence steric hindrance arising from the interactions between the columns promotes the ordering of the dipole moments in a particular direction. The spontaneous polarization in the broken layer columnar phase can be either perpendicular to the columnar axis (the B1 phase with *p2mg* symmetry) or parallel to the columnar axis in the B_{1Rev} type phases. The lattice parameter 'a' provides the number of molecules in the lattice while 'b' corresponds to the molecular length. The density modulations are formed in the plane parallel to the Ps vector in the frequently observed B_1 phase. The B_1 phase possess an antiferroelectric crystallographic unit cell and is non switchable. However on the basis of XRD and electrooptical investigations in some compounds exhibiting switchable phases which are termed as B_{1Rev} phase [51], the density modulations are found to be formed in the plane perpendicular to the Ps vector. The B_{1Rev} type phase with tilted and non-tilted smectic layers is also reported. Based on the molecular structural arrangement the B_{1Rev} phase can be further categorised as orthogonal B_{1Rev} and $B_{1RevTilted}$ phases. In orthogonal B_{1Rev} phases the molecules can rotate around their long axis without destroying the columnar phase structure. However in the switching process of $B_{1RevTilted}$ phases the rotation of molecules occurs either on the tilt cone around the layer normal or around the long molecular axis. In summary the possible structural models of the PMLU and columnar (B_7/$B_{1RevTilt}$) phases had been described [10-12] and 2D modulated structures of the bent core liquid crystalline phases had been proposed based on the reported experimental data. The basic labelled nomenclature [10-12] is SmCPU where SmCP abbreviates the tilted polar molecules in smectic layers and U denotes layer undulation or the breaking of translational symmetry within the plane of the layers by the formation of undulations. Further they had been divided into four classes of $SmCPU_{A,}$ $SmCPU_{S,}$ $SmCPU_S{}'$ and $SmCPU_A{}'$ phases. The first three categories had been observed but the last $SmCPU_A{}'$phase is yet to be observed. This classification enabled to account for only few reported compounds exhibiting B_7 (PMLU) and $B_{1RevTilt}$ phase structures since these phases have the same basic structure and common origin of modulation viz., splay of polarization.

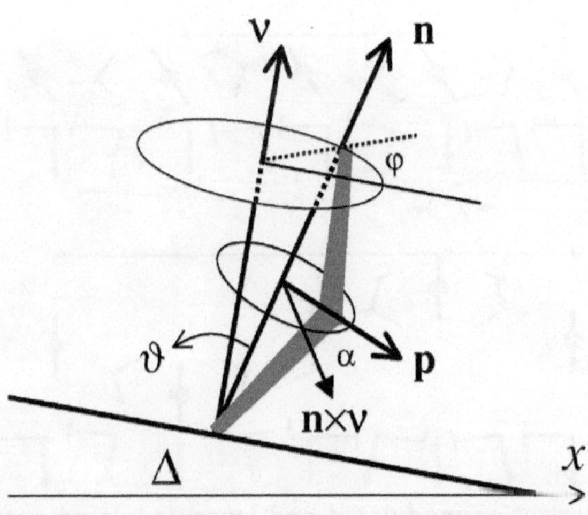

Figure 4.13: The local arrangement of the constituent molecules is described by the nematic director **n**, by the polar parameter **p**, and by the smectic layer

normal **v**. The orientation of these three vectors is described by the cone angle **ϑ**, the tilt of the director with respect to the smectic layer normal, the smectic layer tilt **Δ**, the director position on the cone **φ**, and the angle **α** that describes the tilt of polarization from the direction perpendicular to the tilt plane, the plane determined by the director **n** and the smectic layer normal **v**.

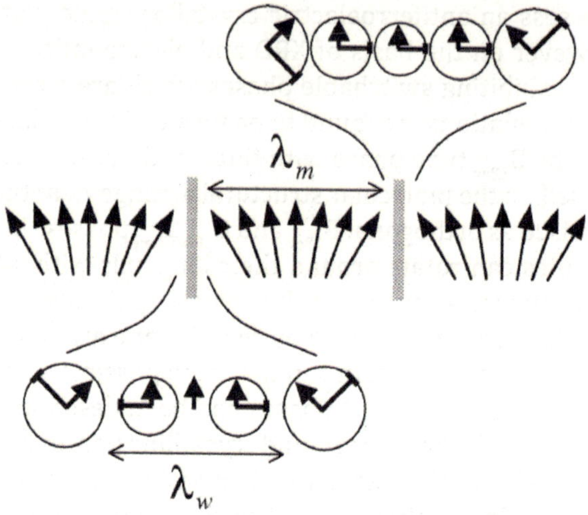

Figure 4.14: In a ferroelectric domain stripes of the preferred polarization splay are divided by walls. The neighbouring stripes can have the same or the opposite chirality. The structure of the wall over which chirality switches is shown in bottom of the figure. The structure of the wall with no chirality switch is shown above. The figure presents the top view of the undulated layer, and at the wall of width λ_w there is the top of a hill or the bottom of a valley. The radius of the circle is a measure for the cone angle, the arrows present the polar parameter, and the lines with bars present the director position on the cone.

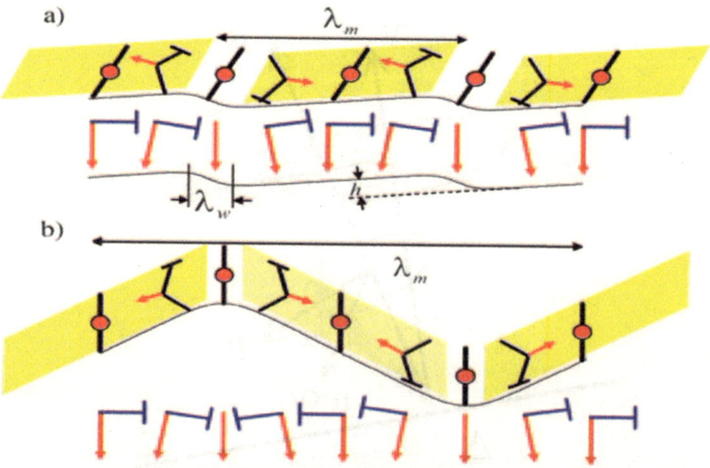

Figure 4.15: Polarization-modulated and layer-undulated structures in bent-core liquid crystals. a) Synclinic layer (block) structure; b) anticlinic layer (block)

structure. Side views and top views of the layers are shown. Arrows represent polarization direction, and lines (⊣) the director position in space, their length being a measure of the magnitude of $\boldsymbol{\vartheta}$. In the side view the ending ⊣ represents the part of the director that is tilted towards the observer. $\boldsymbol{\lambda}_m$ is the modulation wavelength, $\boldsymbol{\lambda}_w$ is the width of the domain wall (the wall between two regions of favourable polarization splay). *h* is the longitudinal shift of the synclinically tilted blocks.

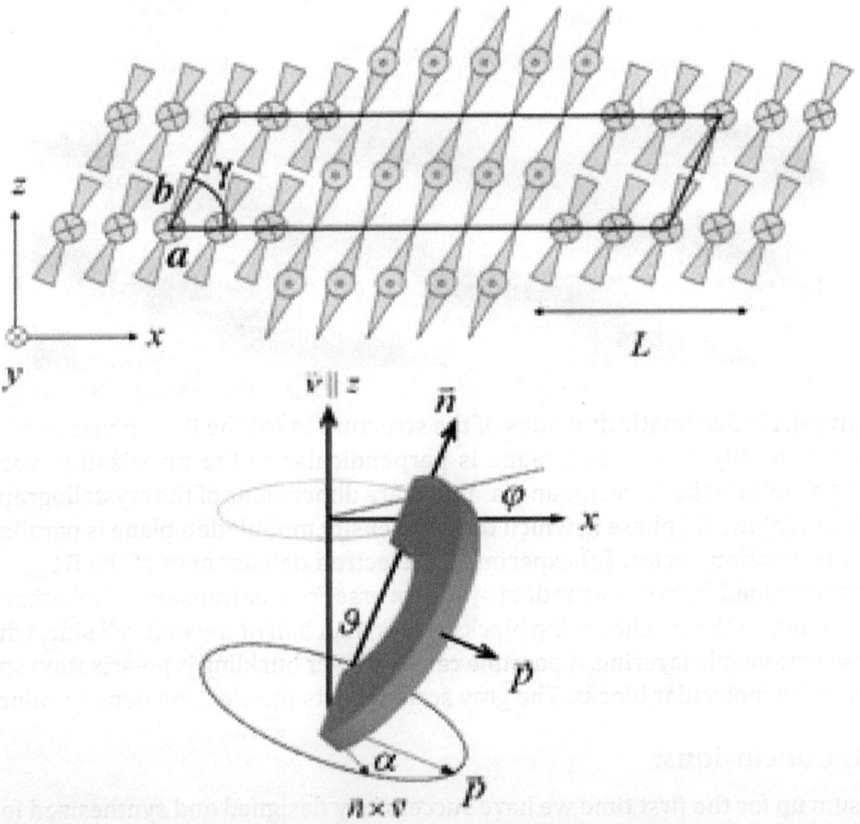

Figure 4.16: Schematic drawings of the $B1_{RevTilted}$ phase structure; L is the width of the column cross-section. The column axes are parallel to the y direction and the layer normal (**v**) is along the z direction. The structure is polarization- modulated along the x direction. Orientation of a single banana molecule in the $B1_{RevTilted}$ phase: the polarization unit vector **p** is perpendicular to the long molecular axis, and the director **ñ** is parallel to the long molecular axis. The orientation of the **p** and **ñ** vectors is determined by the cone angle $\boldsymbol{\vartheta}$, the azimuthal angle $\boldsymbol{\varphi}$, which describes the position of the director **ñ** on the cone, and the angle $\boldsymbol{\alpha}$, which describes the position of the polarization unit vector **p** with respect to the tilt plane normal defined by **ñ** X **v**.

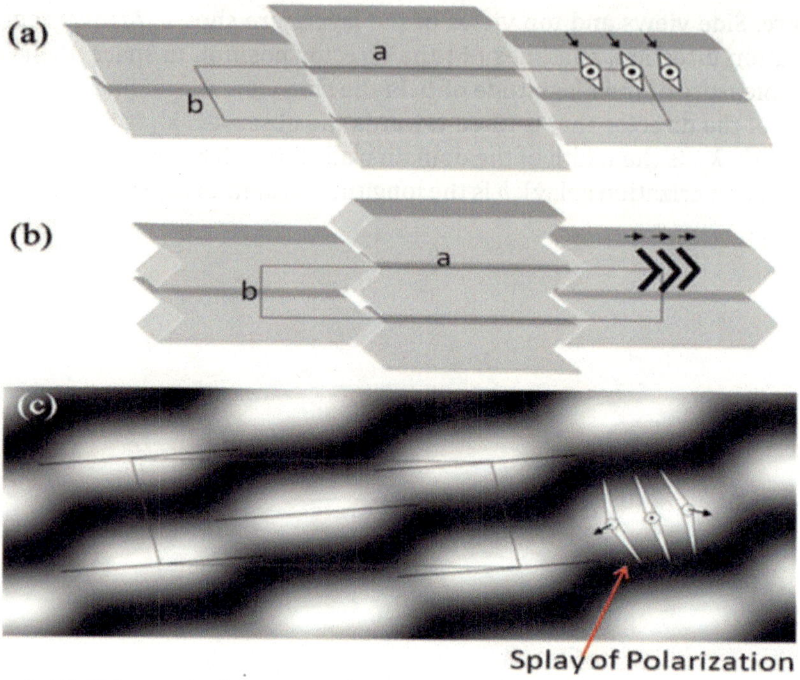

Figure 4.17: Schematic drawings of the structure in (a) the $B1_{Rev}$ phase in which the 2-D density modulation plane is perpendicular to the polarization vector, arrows indicate the Ps vector and 'a' and 'b' are dimensions of the crystallographic unit cell (b) the B1 phase in which the 2-D density modulation plane is parallel to the polarization vector; (c) experimental electron density map of the $B1_{Rev}$-type phase obtained from the X-ray data by the reverse Fourier transform. Note that the shift between the neighbouring blocks is less than half of the unit cell side, which causes the visible layering. A possible cause of layer buckling is polarization splay within the molecular blocks. The gray scale reflects the electron density value.

4.4. Conclusions:

To sum up for the first time we have successfully designed and synthesized four-ring compounds exhibiting 2D polarization splay modulated layer undulated banana mesomorphic phases. The four ring fluorescent systems are promising class of compounds for applications. The novel central core introduced here yield photo-luminescent bent core molecules as well as a new family exhibiting $B1_{RevTilted}$ phases. The compounds exhibit banana mesomorphism possessing $B1_{RevTilted}$ phases, which, when exposed to a sufficiently strong electric field, exhibit the ferroelectric $\mathbf{SmC_SP_F}$ structure. The influence of the orientation of the ester linkage at the molecular bend on the mesomorphic behaviour, chiral characteristics, and phase structural arrangement has been probed. The strong photoluminescence of these molecules opens the door for the structural study of the textures of the $B1_{RevTilted}$ phase by fluorescence confocal polarizing microscopy.

References:

1. S. W. Choi, M. Zennyoji, Y. Takanishi, H. Takezoe, T. Niori and J. Watanabe, *Mol. Cryst. Liq. Cryst.*, 1999, **328**, 185–192.
2. M. Hird, J. W. Goodby, N. Gough and K. J. Toyne, *J. Mater. Chem.*, 2001, **11**, 2732–2742.
3. C. V. Yelamaggad, U. S. Hiremath, S. A. Nagamani, D. S. S. Rao and S. K. Prasad, *J. Mater. Chem.*, 2001, **11**, 1818–1822.
4. D. M. Walaba, E. Korblova, R. Shao and N. A. Clark, *J. Mater. Chem.*, 2001, **11**, 2743–2747.
5. R. Achten, A. Koudjs, Z. Karczmarzyk, A. T. M. Marcelis and E. J. R. Sudholter, *Liq. Cryst.*, 2004, **31**, 215–227.
6. M. Sepelj, A. Lesac, U. Baumeister, S. Diele, D. W. Bruce and Z. Hamersak, *Chem. Mater.*, 2006, **18**, 2050–2058.
7. D. Shen, S. Diele, I.Wirth, and C. Tschierske, *Chem. Commun.*1998, 2573-74.
8. D. Shen, A. Pegenau, S. Diele, I.Wirth, and C. Tschierske, *J. Chem. Am. Soc.*,2000, **122**, 1593-1601.
9. R. Deb, R. K. Nath, M. K. Paul, N. V. S. Rao, F. Tuluri, Y. Shen, R. Shao, D. Chen, C. Zhu, Ivan I. Smalyukh and Noel A. Clark, *J. Mater. Chem.*, 2010, **20**, 7332–7336.
10. G. Pelzl, S. Diele, A. Jákli, C. Lischka, I. Wirth, and W. Weissflog, *Liq. Cryst.*, 1999, **26**, 135.
11. D. A. Coleman, J. Fernsler, N. Chattham, M. Nakata, Y. Takanishi, E. K€orblova, D. R. Link, R. F. Shao, W. G. Jang, J. E. Maclennan, O. M. Monval, C. Boyer, W. Weissflog, G. Pelzl, L. C. Chien, D. M. Walba, J. Zasadzinski, J. Watanabe, H. Takezoe and N. A. Clark, *Science*, 2003, **301**, 1204–1211.
12. H. N. S. Murthy, B. K. Sadashiva Liq. Cryst., 2003, **30**, 1051.
13. (a) D. A. Coleman, C. D. Jones, M. Nakata, N. A. Clark, D. M. Walba, W. Weissflog, K. Fodor-Csorba, J. Watanabe, V. Novotna and V. Hamplova, Phys. Rev. E: Stat., Nonlinear, *Soft Matter Phys.*, 2008, **77**, 021703; (b) N. Vaupotic, M. Copic, E. Gorecka and D. Pociecha, *Phys. Rev. Lett.*, 2007, **98**, 247802.
14. P. J. Collings, *Liquid Crystals: Nature's Delicate Phase of Matter*, Princeton University, Princeton, NJ, (1990).
15. (a) E. Thikonov, M. Bertolotti, F. Scudieri, *Appl. Phys.*, 1976, **11**, 35; (b) M. Bertolotti, G. Sansoni, F. Scudieri, *Appl. Opt.*, 1979, **18**, 528.

16. D. Bauman, A. Skibínski, R. Stolarski, *Mol Cryst. Liq. Cryst.*, 1986, **138**, 367.
17. T. Marty'nski, E. Mykowska, D. Bauman, *J. Mol. Struct.*, 1994, **325**, 161.
18. H. J. Coles, G.A. Lester, H. Owen, *Liq. Cryst.*, 1993, **14**, 1039.
19. A. Beer, G. Scherowsky, H. Owen, H. Coles, *Liq. Cryst.*, 1995, **19**, 565.
20. G. H. Heilmeier, L. A. Zanoni, *Appl. Phys. Lett.*, 1968, **13**, 91.
21. T. Urisu, K. Kajiyama, Y. Mizushima, *Appl. Opt.*, 1978, **17**, 2366.
22. E. Wolarz, D. Bauman, *Mol. Cryst. Liq. Cryst.*, 1991, **197**, 1.
23. T. Martynski, K. Fiksinski, H. Moryson, W. Mykowska, E. Wolarz, D. Bauman, *SPIE*, 1995, **2372**, 317.
24. S. Kato, B. Lee, C. Pac, *Liq. Cryst.*, 1997, **22**, 595.
25. R. Subramanian, L. K. Patterson, H. Levanon, *Chem. Phys. Lett.*, 1982, **93**, 578.
26. T. Ikeda, S. Kurihara, S. Tazuke, *J. Phys. Chem.*, 1990, **94**, 6550.
27. N. K. Chudgar, M. K. Parekh, S. S. Madhavrao, H. C. Sharma, *Liq. Cryst.*, 1995, **19**, 807.
28. R. E. Gill, A. Meetsma, A. Hadziioannou, *Adv. Mater.*, 1996, **8**, 212.
29. D. T. Breslin, M. A. Fox, *Chem. Phys. Lett.*, 1995, **239**, 299.
30. J. Simmerer, B. Glüsen, W. Paulus, A. Kettner, P. Schumacher, D. Adam, K.-H. Etzbach, K. Siemensmeyer, J. H. Wendorff, H. Ringsdorf, D. Haarer, *Adv. Mater.*, 1996, **8**, 815.
31. T. Christ, B. Glüsen, A. Greiner, A. Kettner, R. Sander, V. Stümpflen, V. Tsukruk, J. H. Wendorff, *Adv. Mater.*, 1997, **9**, 48.
32. T. Christ, V. Stümpflen, J. H. Wendorff, *Macromol. Rapid Commun.*, 1997, **18**, 93.
33. (a) H. Gallardo, R. Cristiano, A. A. Vieira, R. A. W. Neves Filho, R. M. Srivastava and I. H. Bechtold, *Liq. Cryst.*, 2008, **35**, 57–63; (b) A. A. Vieira, R. Cristiano, A. J. Bortoluzzi and H. Gallardo, *J. Mol. Struct.*, 2008, **875**, 364–371; (c) R. M. Srivastava, R. A. W. Neves Filho, R. Schneider, A. A. Vieira and H. Gallardo, *Liq. Cryst.*, 2008, **35**, 737–742; (d) R. Cristiano, F. Ely and H. Gallardo, *Liq. Cryst.*, 2005, **32**, 15–25; (e) A. A. Vieira, R. Cristiano, F. Ely and H. Gallardo, *Liq. Cryst.*, 2006, **33**, 381–90.
34. F. Camerel, L. Bonardi, M. Schmutz and R. Ziessel, *J. Am. Chem. Soc.*, 2006, **128**, 4548–49.
35. Joshi, M. O. Manasreh, E. A. Davis and B. D. Weaver, *Appl. Phys. Lett.*, 2006, **89**, 111907-1–3.
36. S. Chandrasekhar, B. K. Sadashiva, K. A. Suresh, *Pramana* 1977, **9**, 471.

37. W. Weissflog, I. Letko, S. Diele, G. Pelzl, *Adv. Mater.* 1996, **8**, 76.
38. I. Letko, S. Diele, G. Pelzl, W. Weissflog, *Liq. Cryst.* 1995, **19**, 643.
39. A. I. Smirnova, D. Fazio, E. F. Iglesias, C. G. Hall, D. Guillon, B. Donnio, D. W. Bruce, *Mol. Cryst. Liq. Cryst.*, 2003, **396**, 227.
40. A. C. Ribeiro, B. Heinrich, C. Cruz, H. T. Nguyen, S. Diele, M. W. Schroder, D. Guillon, *Eur. Phys. J. E*, 2003, **10**, 143.
41. N. Vaupotic and M. Copic, *Phys. Rev. E*, 2005, **72**, 031701.
42. N. Vaupotic, *Ferroelectrics*, 2006, **344**, 151.
43. J. Watanabe, T. Niori, T. Sekine, H. Takezoe, *Jpn. J. Appl. Phys.* Part 1 1998, **37**, L139.
44. G. Pelzl, S. Diele, W. Weissflog, *Adv. Mater.* 1999, **11**, 707.
45. J. P. Bedel, J. C. Rouillon, J. P. Marcerou, M. Laguerre, H. T. Nguyen, M. F. Achard, *Liq. Cryst.*, 2000, **27**, 1411.
46. J. Szydlowska, J. Mieczkowski, J. Matraszek, D. W. Bruce, E. Gorecka, D. Pociecha, D. Guillon, *Phys. Rev. E*, 2003, **67**, 031702.
47. J. Mieczkowski, K. Gomola, J. Koseska, D. Pociecha, J. Szydlowska, E. Gorecka, *J. Mater. Chem.*, 2003, **13**, 2132.
48. D. Pociecha, N. Vaupotic, E. Gorecka, J.Mieczkowski, K. Gomola, *J. Mater. Chem.*, 2008, **18**, 881.
49. K. Pelz, W. Weissflog, U. Baumeister, S. Diele, *Liq. Cryst.*, 2003, **30**, 1151.
50. C. Keith, R. Amaranatha Reddy, U. Baumeister, C. Tschierske, *J. Am. Chem. Soc.*, 2004, **126**, 14312.
51. E. Gorecka, N. Vaupotic, D. Pociecha, M. Cepic, J. Mieczkowski, *ChemPhysChem*, 2005,**6**, 1087-1093

CHAPTER 5

Effect of Polar Substituents on Mesomorphism

Our study of examining the effect of polar substituents on the central bent core unit of achiral four ring banana shaped molecules are detailed in this chapter. The phases exhibited by these compounds were confirmed by POM, DSC and XRD techniques.

5. Effect of polar substituents on mesomorphism of unsymmetrical four-ring bent-core compounds:

In the previous chapter it is successfully demonstrated that new asymmetric four-ring bent-core compounds exhibiting banana mesomorphism viz., 2D-polarization modulated layer undulated smectic phases ($B1_{RevTilt}$ phases). The majority of the reported four-ring bent-core compounds did not exhibit mesomorphism [1-5]. HoweverHird et al reported fluoro substituted bent–core compounds exhibiting monotropic nematic phase with chiral domains. The earlier reported four-ring molecules with a biphenyl moiety in the wing possessing fluoro substituents on the central ring viz. [(S)-1-methylheptyl-4-[2,3,4-trifluoro-5-(4/-dodecyloxy-biphenyl-4-ylcarbonyloxy)-phenylcarbonyloxy] benzoate] exhibited conventional SmA, ferro-, ferri-, and antiferroelectric SmC phases resulting from the molecular chirality and not due to the bent shape [6]. Weissflog et al recently reported asymmetric achiral four-ring bent-core mesogens derived from N-benzoylpiperazine and only few of them found to exhibit only calamitic phases. The introduction of a substituent in the side wings of the molecules also does not promote the mesomorphism [7]. They reported polar structures under the influence of electric field in these four-ring bent-core compounds exhibiting calamitic type smectic phases. Moreover they also exhibited switchable smectic I phase the first example of non-chiral ferroelectric SmI phase ($SmIP_F$), which is attributed due to bent shape of the molecules. None of these four-ring compounds are found to spontaneously exhibit typical banana phases (SmAP, SmCP) because of weak steric interactions between the aromatic cores. Kang et al [8] reported a four-ring oxadiazole derivative exhibiting nematic-SmC-Bx phase variant. The Bx phase is found to be uniaxial, exhibited randomly distributed domains of opposite handedness and under the influence of electric field transformed to a homochiral SmC_AP_F phase with antiferroelectric response.

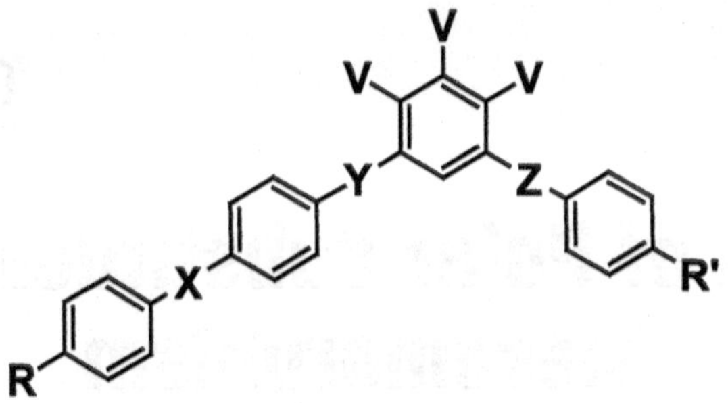

R	X	Y	Z	V	R/	Mesophase	Ref.
$C_{12}H_{25}O$	COO	CH=CH	OOC	H	$OC_{12}H_{25}$	Cr 108 I	1
NC-	--	OOC	OOC	H	$OC_{10}H_{21}$	Cr 141 I	2
$C_7H_{15}O$	COO	COO	COO	H	OC_8H_{17}	Cr 130 I	3
$C_{10}H_{21}O$	CH=CHCOO	CH=N	OOC	H	OC_8H_{17}	Cr 96 I	
H	N=N	N=N	OOC	F	$OC_{12}H_{25}$	Cr 107 I	4
$C_{10}H_{21}O$	--	OOC	OOC	F	$OC_{12}H_{25}$	Cr 88.5 I	5
$C_{12}H_{25}O$	--	COO	OOC	F	$OC_{12}H_{25}$	Cr 70 (N 41) I	5

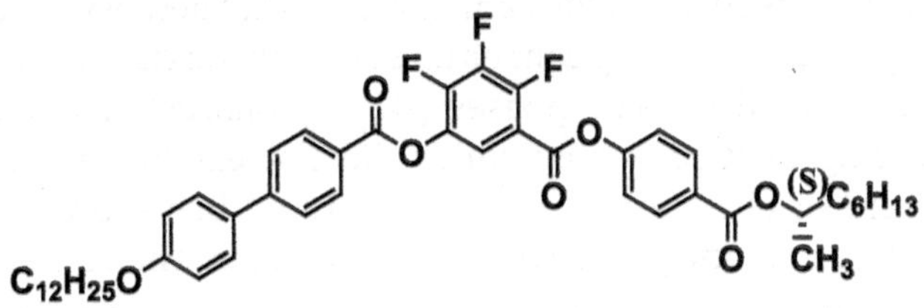

C 99.5 $S_C{}^*{}_{anti}$ 137.5 $S_C{}^*{}_{ferri}$ 141.0 $S_C{}^*{}_{ferro}$ 166.0 $S_A{}^*$ 213.5 I — Ref. 6

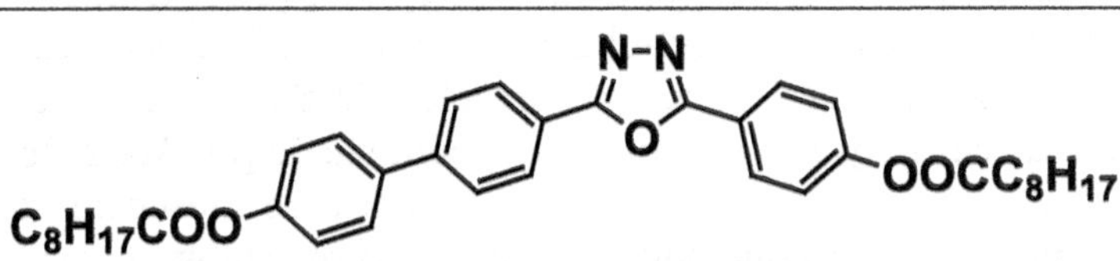

Is 185 N 181 SmC 160 Bx 69 Cr — Ref. 7

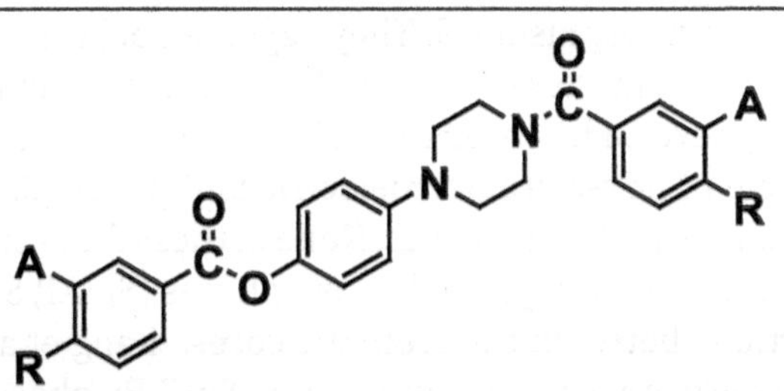

R = $C_nH_{2n+1}O$	**A**	**Mesophase**	Ref.
n = 4, 6,	H	Nonmesomorphic	8

n = 7, 8, 9	H	Monotropic SmA	
n = 10, 11, 12, 14, 16	H	Monotropic SmI- SmC -SmA	
n = 12	F/Cl/Br	Nonmesomorphic	
R = $C_{10}H_{21}NH$	H	Nonmesomorphic	
R = $C_{12}H_{25}$	H	Monotropic SmA	
R = $C_{10}H_{21}OOC$	H	Monotropic SmA	
R=$C_{12}H_{25}O$ 2,6-naphthylene instead of terminal phenyl ring	H	Nonmesomorphic	

Figure 5.1: Molecular structures of reported four ring compounds

The liquid crystalline phase behaviour of bent or banana-shaped compounds is dependent upon several factors, namely the size of the molecule and the number of aromatic rings, the position as well as the magnitude of the bent angle, the size, number, position and nature of the substituents, the nature and direction of linkage groups and the length and nature of the terminal alkyl chains. In general, any minor change in these structural elements leads to drastic changes in the phase behaviour. One of the important aspects is the nature and size of the substituent in the central core, which largely influences the mesophase behaviour [9-13]. Several research groups studied the influence of the lateral substituents on the mesophase behaviour either in the central core or the outer rings [9-19]. However, the introduction of a small substituent into the central 1,3-phenylene ring leads to the possibility of synthesising new mesogens with novel banana phases [16-21]. Moreover, it was found that mesophase behaviour is much more strongly influenced by substituents at the central core than by these at the outer ring [22-25]. Rich variety of banana phases was discovered by introducing lateral substituents in the central ring. In case of lateral substitution three major factors are involved depending on the size and polarity of the substituents namely inter- and intra molecular forces of attraction, repulsion of the substituents and the position of the substitution. These factor influence molecular conformation. Molecular conformation effects on molecular packing and vice versa. Consequently, lateral substitution has an impact on the liquid crystalline state, i.e. on the liquid crystalline behaviour. Systematic study of banana substance classes one by one, may advance understanding of this subject.

In this and next chapters unsymmetrical four ring banana shaped compounds substituted with Fluoro (F), Chloro (Cl) and Methyl (CH_3) at different positions of the aromatic ring of the central core will be discussed. This chapter is divided into three sections based on the substituents e.g. the first section is on the effect of fluoro substituent, the second and third section on the effect of chloro substituents at different positions.

Modified Structures of central bent core unit with a substituent

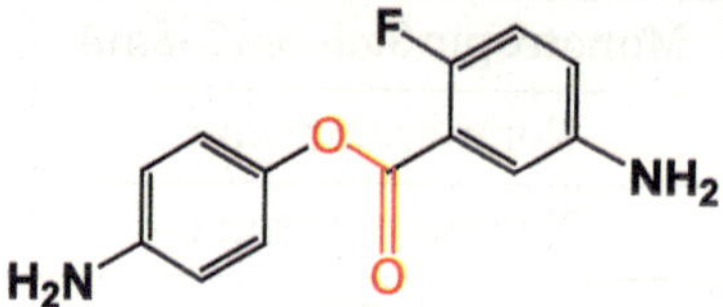

5-3a

4′-aminophenyl -5-amino-2-fluorobenzoate

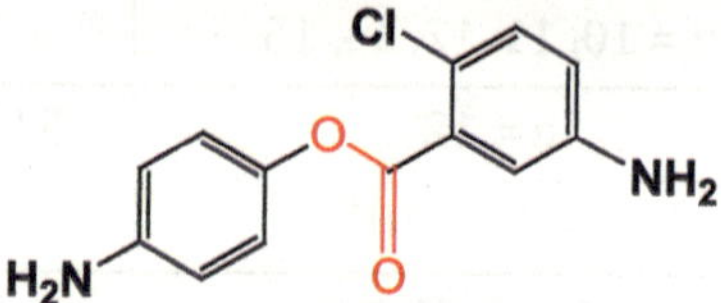

5-3b

4′-aminophenyl -5-amino-2-chlorobenzoate

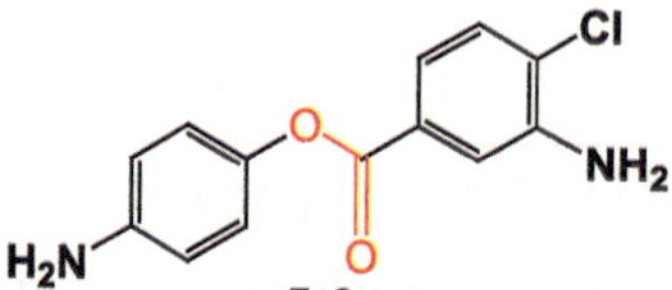

5-3c

4′-aminophenyl -3-amino-4-chlorobenzoate

Designed compounds

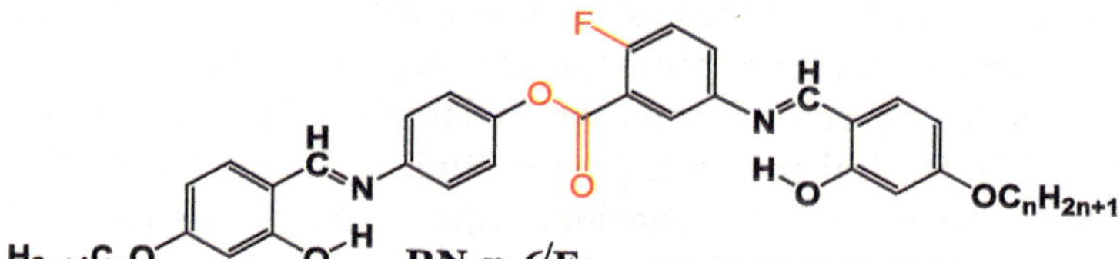

RN-n-6′F: n = 8, 9, 10, 11, 12, 13, 14, 16

*[4-(N-4′-n-alkyloxysalicylidene)-aminophenyl]-[6-fluoro-3-(N-4′-n-alkyloxysalicylideneamino)-benzoate***RN-n-6′F**

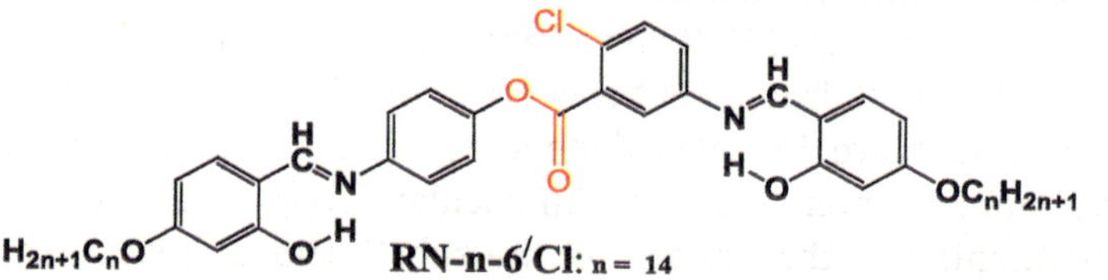

RN-n-6′Cl: n = 14

*[4-(N-4′-n-alkyloxysalicylidene)-aminophenyl]-[6-chloro-3-(N-4′-n-alkyloxysalicylideneamino)-benzoate***RN-n-6′Cl**

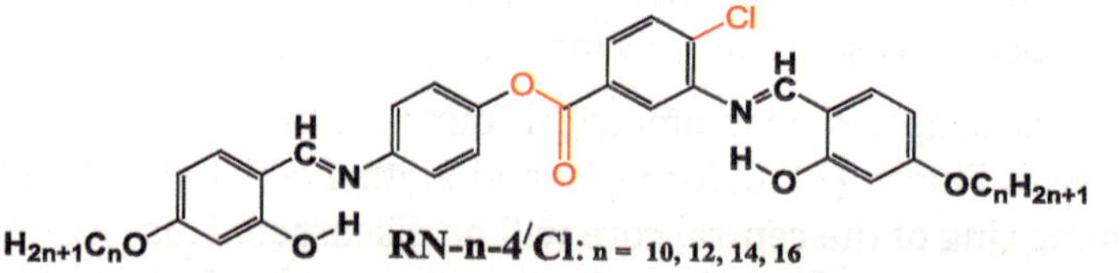

RN-n-4′Cl: n = 10, 12, 14, 16

*[4-(N-4′-n-alkyloxysalicylidene)-aminophenyl]-[4-chloro-3-(N-4′-n-alkyloxysalicylideneamino)-benzoate***RN-n-4′Cl**

Figure 5.2: Modified Structures of central bent core unit with a substituent and designed colpounds

5.1. Effect of Fluoro substituents:

The fluoro substituent in organic compounds is regarded as so interesting because of the combination of polar and steric effects, and the great strength of the C–F bond which confers stability on fluoro-substituted compounds. Fluorine has the highest electro-negativity of all the elements (3.98), and hence as a substituent confers a high dipole moment on the C–F bond. In an aliphatic or alicyclic environment the dipole moment is relatively large, e.g., 1.85 D in fluoromethane, however, in an aromatic environment the mesomeric effect causes a reduction in dipole moment, e.g., 1.50 D for fluorobenzene. Despite the high polarity, the fluoro substituent has a low polarizability which confers low intermolecular dispersion interactions. The fluoro substituent is the smallest, after hydrogen, of all possible substituents, and like hydrogen it is monoatomic. So although a fluoro substituent obviously causes a steric effect, the size influence is not too drastic, which enables it to be usefully incorporated into parent molecules for beneficial modification of properties. Lateral substituents, particularly fluoro, are frequently employed in liquid crystal structures to modify melting point, liquid crystal transition temperatures and mesophase morphology and to modify the physical properties of liquid crystals to enable their use in applications. Fluoro substituents have been so successfully and usefully incorporated into liquid crystal molecules because of the combination of small size and high polarity [26-28] and because the high strength of the C–F bond confers excellent stability. The relatively small size of a fluoro substituent means that it does not unduly alter the necessary stylized structure, which helps to maintain the existing liquid crystalline nature of the compound. However, the important attributes of the fluoro substituent discussed above, ensures that subtle, but significant, modifications are frequently encountered in respect of melting point, mesophase morphology, transitions temperatures, and the many essential physical properties of liquid crystals, such as optical, dielectric, and visco-elastic properties. Such alterations to the attributes of liquid crystals are of great significance in terms of fundamental academic structure–property relationships, and crucial to the development of commercially-successful liquid crystal displays. The structure of this series of compound is shown in **figure 5.2.**

5.1.1. Synthesis:

The starting material in the present study 4-n-alkyloxysalicylaldehyde was prepared by Williamson etherification of 2,4-dihydroxybenzaldehyde with appropriate n-alkyl bromide. The central bent core moiety 4/-aminophenyl-5-amino-2-fluorobenzoate (**5-3a**) was prepared as follows. 2-Fluoro-5-nitrobenzoic acid was converted into its acyl chloride followed by condensation with 4-nitrophenol using phase transfer reaction. The resultant 4/-nitrophenyl-2-fluoro-5-nitrobenzoate was subjected to 10% Pd-C catalysed reduction to get the desired di-amino bent-core compound. The condensation of 4-n-alkyloxy salicylaldehyde with 4-aminophenyl-5-amino-2-fluorobenzoate in presence of a few drops of glacial acetic acid yielded the target bent shaped compounds [4-(N-4/-

n-decyloxysalicylidene)-aminophenyl]-[5-(N-4′-n-decyloxysalicylideneamino)-2-fluorobenzoate (**RN-n-6′F, n = 10**). To avoid the formation of side products, the precipitated compounds were filtered when the solution was hot to yield the pure compounds. The compounds were further recrystallized repeatedly to get the pure samples. The formation of all of the compounds was confirmed by ^{1}H-NMR and IR spectroscopy and the purity was established by elemental analysis. The other homologues with n = 8, 9, 11, 12, 13, 14 and 16 were also synthesized following the same procedure and characterized. The details of the experimental procedures along with the spectroscopic data for all the homologous series of compounds are presented in the experimental part of **Chapter 3**. The liquid-crystalline behaviour of the synthesised compounds had been investigated by optical microscopy and differential scanning calorimetry (DSC).

5.1.2. Mesomorphic properties:

The transition temperatures, enthalpies and entropies associated with the phase transitions of all the compounds of the homologous series **RN-n-6′F** (n = 8 to 14 and 15) as a function of number of the carbon atoms in the terminal alkyl chains obtained from DSC at a scan rate of 5^{0}C min^{-1} in the second heating and first cooling scans are presented in **Table 5.1.** All the compounds of the homologous series **RN-n-6′F** exhibitbananaliquid crystalline behaviour over a moderate temperature region. The differential scanning calorimetric thermogram of representative compounds **RN-14-6′F** and **RN-8-6′F** are presented in **Figure 5.3. RN-14-6′F** exhibited two transitions in the heating cycle at 126.63^{0}C (ΔH = 31.58 kJ/mol, ΔS = 79.02 J/K/mol), melting point and 144.45^{0}C (ΔH = 56.95 kJ/mol, ΔS = 136.42 J/K/mol) clearing point. In the cooling cycle also it exhibited enantiotropic phase transitions at 138.37^{0}C (ΔH = 19.13 kJ/mol, ΔS = 46.50 J/K/mol), 99.75^{0}C (ΔH = 19.80 kJ/mol, ΔS = 53.11 J/K/mol).

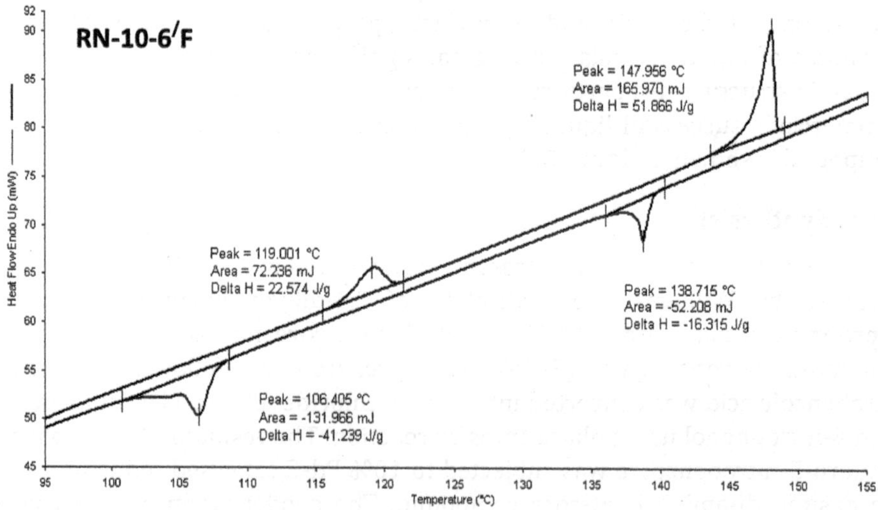

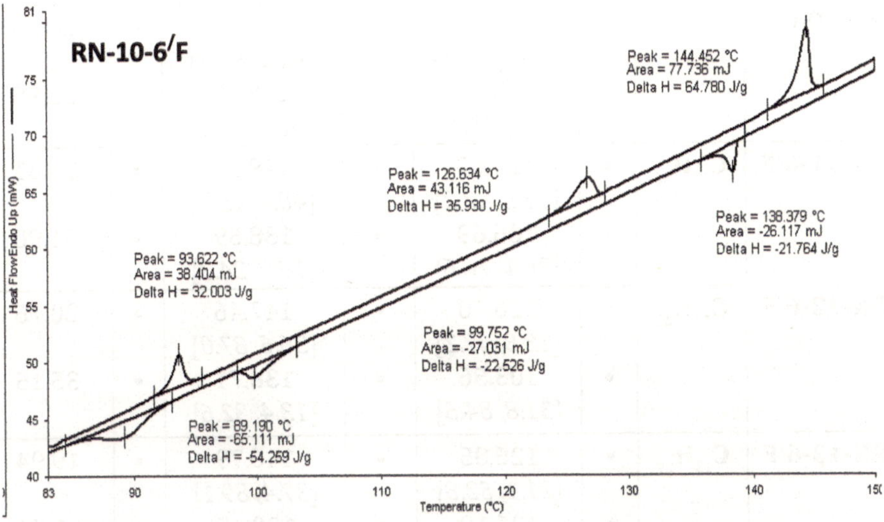

Figure 5.3: DSC thermogram of **RN-14-6/F** and **RN-10-6/F.**

A plot of phase transition temperatures as a function of number of carbons in the end alkoxy chain is depicted in **Figure 5.4**. All the compounds were found to exhibit one mesophase only exhibiting almost same texture. The mesophase range increases with the increasing *n*-alkoxy chain length initially and for the number of carbon atoms 10 and more the mesomorphic range is almost constant (**Figure 5.5**). The clearing transition temperatures for the mesophase in a homologous series follow a smooth curve.

Table 5.1: Transition Temperatures (^{0}C), Enthalpies (*Italic*,kJ/mol) and entropies (*Italic*, J/mol/K) of homologous series **RN-n-6/F**.

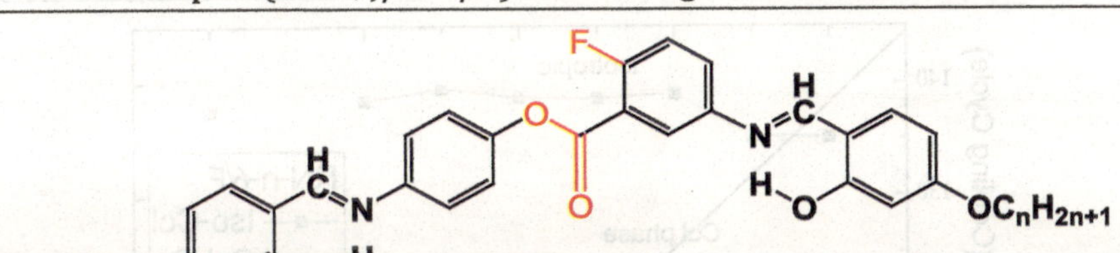

RN-n-6/F: **n = 8, 9, 10, 11, 12, 13, 14, 16**

Compound	R	Cr	**Heating** **Cooling**	Col$_r$	**Heating** **Cooling**	I	ΔT
RN-8-6/F	C_8H_{17}	•	135.76 *[20.0, 48.9]*	•	156.09 *[48.2, 112.2]*	•	20.33
		•	118.50 *[24.3, 62.1]*	•	135.14 *[16.5, 40.4]*	•	16.64
RN-9-6/F	C_9H_{19}	•	131.33 *[14.8, 36.6]*	•	152.62 *[36.7, 86.3]*	•	21.29
		•	112.92 *[31.4, 81.4]*	•	135.55 *[7.2, 17.6]*	•	22.63

RN-10-6/F	$C_{10}H_{21}$	• •	127.13 *[20.9, 52.3]* 105.99 *[9.5, 25.1]*	• •	147.63 *[41.1, 97.7]* 138.97 *[14.1, 34.2]*	• •	20.50 32.98
RN-11-6/F	$C_{11}H_{23}$	• •	124.67 *[16.8, 42.2]* 106.69 *[29.1, 76.6]*	• •	149.28 *[40.4, 95.7]* 138.59 *[10.0, 24.3]*	• •	24.61 31.90
RN-12-6/F	$C_{12}H_{25}$	• •	126.70 *[18.8, 47.2]* 103.56 *[31.8, 84.5]*	• •	147.46 *[36.6, 87.0]* 138.71 *[13.4, 32.6]*	• •	20.76 35.15
RN-13-6/F	$C_{13}H_{27}$	• •	126.85 *[21.1, 52.8]* 106.19 *[12.1, 31.9]*	• •	146.79 *[37.4, 89.1]* 139.45 *[12.3,29.8]*	• •	19.94 33.26
RN-14-6/F	$C_{14}H_{29}$	• •	126.63 *[31.6, 79.0]* 99.75 *[19.8, 53.1]*	• •	144.45 *[56.5, 41.7]* 138.37 *[19.1, 46.5]*	• •	17.82 38.62
RN-16-6/F	$C_{16}H_{33}$	• •	111.73 *[76.0, 197.5]* 101.68 *[27.1, 72.5]*	• •	141.95 *[50.8, 122.4]* 137.47 *[16.2, 39.6]*	• •	30.22 35.79

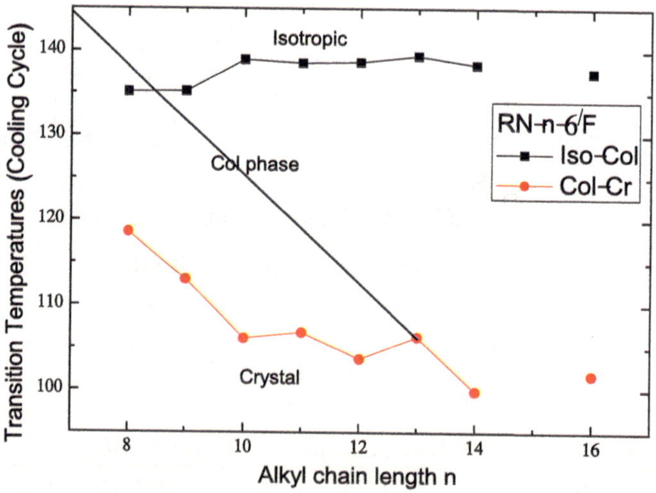

Figure 5.4: Plot of phase transition temperatures as a function of number of carbons in the end alkoxy chain for the compounds of series **RN-n-6/F** in the cooling cycle.

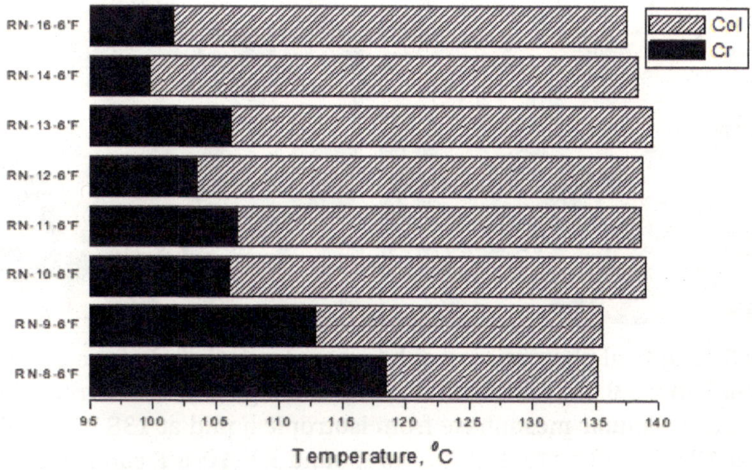

Figure 5.5: Influence of the number of carbon atom at the end alkoxy chain upon the liquid crystalline properties of **RN-n-6/F** series.

The compounds of the homologous series exhibit only one mesophase. The thin film of these samples examined by polarizing optical microscopy highly spherulitic growth was observed from the isotropic melt upon cooling. The typical microphotographs of the textures exhibited by different compounds are shown in **Figure 5.6**. On slow cooling the isotropic melt of compound **RN-14-6/F**, B1 phase grows as dendritic nuclei which coalesce into a typical mosaic texture with large domains of uniform birefringence characterising the phase as columnar phase (**Figure 5.6c**). The texture of the mesophase (**Figure 5.6a, b**) is dominated by defects growing from spherulitic nuclei into banana-leaf or lancet-like defects. The observed POM textures resemble the textures of B1 phase reported in other bent core compounds [11, 29-34]. The XRD studies also confirmed the phase as B1. A comparison of the mesomorphic behaviour of these compounds with that of parent unsubstituted compounds reveals that a) a strong super cooling in the Isotropic-B1 phase transition temperatures, b) rectangular columnar phases instead of 2D modulated reversible tilted phases.

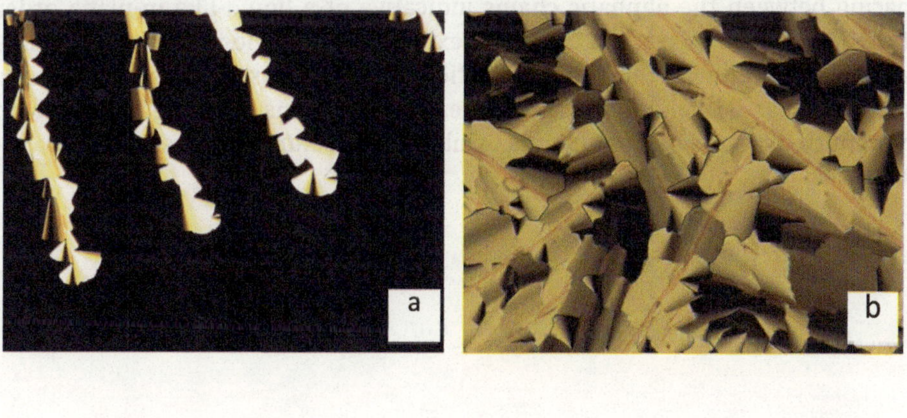

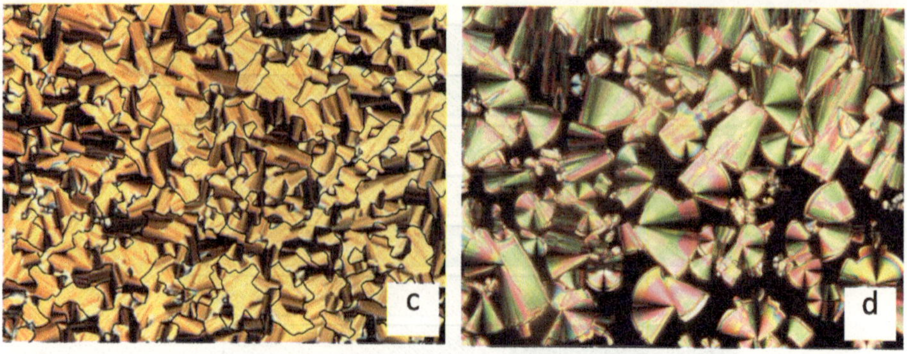

Figure 5.6: Optical photomicrograph (crossed polarizers) of the Col_r phase of **RN-14-6/F** in a 5.0 μm cell as obtained by cooling from the isotropic liquid a) growing of columnar mesophase from isotropic liquid at 138 ^{0}C b) columnar phase at 125.5 ^{0}C c) texture of the compound **RN-10-6/F** contained between glass plate and cover slide at 120^{0}C. d) Columnar phase of **RN-11-6/F** at 132 ^{0}C

5.1.3. X-Ray Studies:

To confirm the mesophase structures X-ray diffraction studies are performed on a representative compound belonging to the homologous series **RN-*n*-6/F**. X-Ray measurements were performed on unaligned (powder) samples in a 0.5 mm diameter Lindeman capillary tube (sealed at both ends) with Cu Kα (λ = 1.5418Å) radiation using the P Analytical X'Pert PROMP X-ray diffractometer consisting of a focusing elliptical mirror and a fast resolution detector (PIXCEL). A typical diffraction pattern obtained for the compound **RN-10-6/F** at 134^{0}C is shown in **Figure 5.7.** The diffraction pattern indicated a strong reflection at $2\theta = 1.95^0$ corresponding to 45.2Å. This corresponds to the molecular length n which is slightly lower than the estimated molecular length of 48.7 Å (**Figure 5.8**) in all trans conformation indicated a tilted layer with a tilt angle 21^0. Another broad diffuse peak at $2\theta = 5.776^0$ corresponding to 15.3 Å in conjunction with the peak at 45.2 Å suggest a 2D columnar phase. In addition to these reflections a broad diffuse wide angle reflection at $2\theta = 19.21^0$ corresponding to 4.62 Å reflects the spacing between the aliphatic chains indicative of a liquid-like in-plane order of the fluid phase. Based on the POM texture, X-ray studies and large enthalpy values associated with the phase transition the mesophase in these compounds is identified as B1 columnar phase. These observations are in agreement with the reported results on compounds exhibiting B1 columnar phases [18, 30, 35-36].

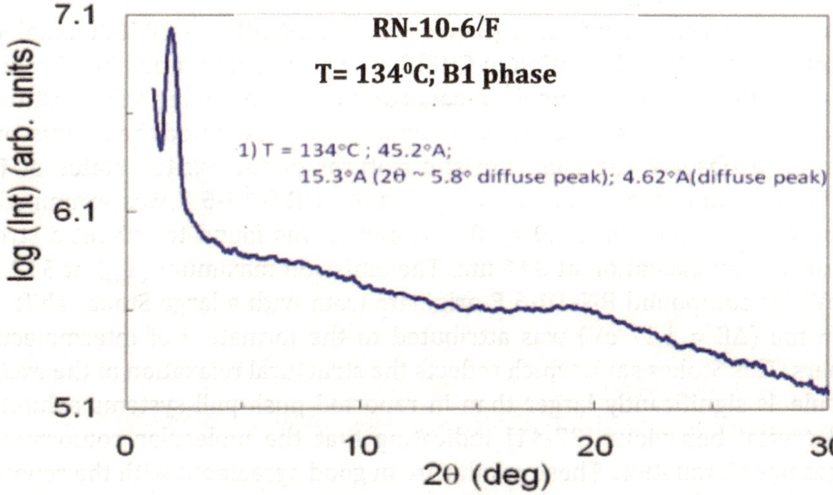

Figure 5.7: X-Ray angular intensity profile obtained for the B1 mesophase of compound **RN-10-6/F** at 134 °C diffraction spectrum,

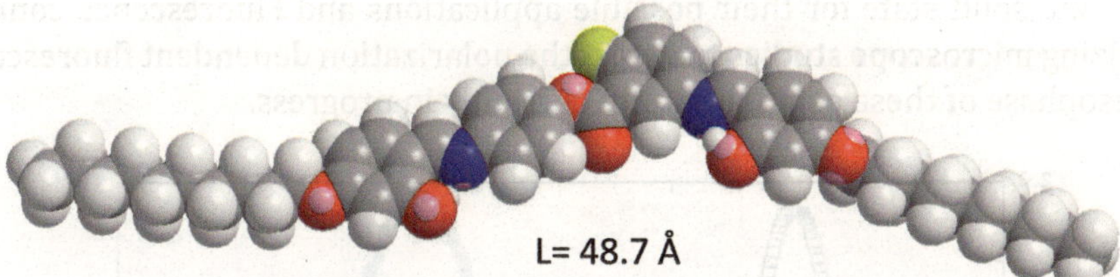

Figure 5.8: Space filling model of the molecule **RN-10-6/F** in all trans conformation. The molecular length of the compound is found to be 48.7Å.

5.1.4. UV-Visible and Fluorescence studies:

As already described in previous chapter the emission properties of organic liquid crystals (LCs) are currently an active area of research because of their potential applications in the area of optoelectronics. Emissive liquid crystals (LCs) are intriguing systems for materials research because they couple molecular self-assembly with intrinsic light generation capabilities. The importance of the substituent on the central core plays an important role in the modification of electronic characteristics of the molecules. Hence the absorption and emission properties of the compounds are studied and the results are presented in this section.

The UV absorption and fluorescence spectroscopic properties of compound **RN-10-6/F** in solution were studied in various solvents of different concentrations to obtain information regarding absorption and emission maxima and the Stokes shift of fluorescence. The UV-Visible and Fluorescence spectra of a representative compound **RN-10-6/F** are shown in **figure 5.9.**The UV-visible spectra in these homologous compounds exhibited strong absorption peaks at ~344 nm with a large molar extinction coefficient (3.60 eV for **RN-10-6/F**, $\varepsilon \sim 107800$ $Lmol^{-1}cm^{-1}$, concentration = 10^{-5} M solution in $CHCl_3$). The strong absorption band reflects the π-π* transition of the highly π-conjugated system having the substituted phenyl benzoate unit as the core. Additionally, we investigated the fluorescence spectrum to observe the characteristic features of the excited states of **RN-10-6/F**. The fluorescence emission spectrum of **RN-10-6/F** was examined in chloroform solution (conc. 1.0 x 10^{-5} M) and it was found to exhibit a strong fluorescence on excitation at 344 nm. The emission maximum (λ_{em}) at 510 nm (2.4 eV) for compound **RN-10-6/F** originate from with a large Stokes shift (λ_S) of 166 nm (ΔE = 1.17 eV) was attributed to the formation of intermolecular excimers. This Stokes shift, which reflects the structural relaxation of the excited molecule, is significantly larger than in reported push-pull systems exhibiting liquid crystal behaviour [37-41] indicating that the molecular conformation changes upon excitation. These results are in good agreement with the reported results of molecular J-aggregates [42] in which the excitonic energy is delocalized as a result of intermolecular coupling within the head-to-tail arrangement of the molecules in the solution. The optical band gap investigations of thin films in LC phase and solid state for their possible applications and Fluorescence confocal polarizing microscope studies to detect the polarization dependent fluorescence in mesophase of these compounds are currently in progress.

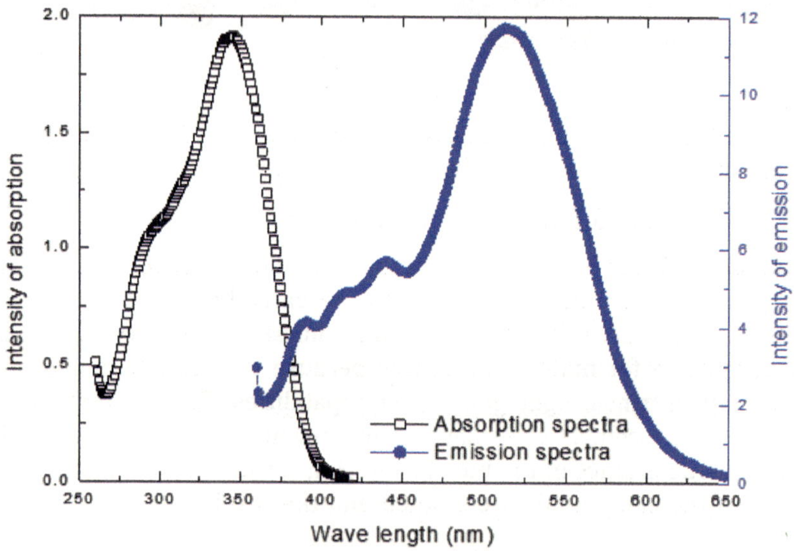

Figure 5.9: UV-Visible and fluorescence spectra of **RN-10-6/F**

5.2. Effect of Chloro substituents:

Chloro substitution is less studied compared to fluoro substitution [43-45]. The size of chlorine atom (atomic radius = 0.97Å) is larger than fluorine atom (atomic radius = 0.57Å). Because of larger atomic size of the chloro substituent than the hydrogen and fluorine substituents, it exerts a steric effect on the organization of molecular structures and in turn mesomorphic characteristics.

5.2.1. Synthesis of [4-(N-4-tetradecyloxysalicylidenemino)phenyl]-[5-(N-4-tetradecyloxysalicylidenemino)-2-chlorobenzoate], RN-14-6/Cl:

The starting material 4-n-tetradecyloxy salicylaldehyde was condensed with the unsymmetric central core 4/-aninophenyl-5-amino-2-chloro benzoate to get the final compound, [4-(N-4-tetradecyloxysalicylidenemino)phenyl]-[5-(N-4-tetradecyloxy-salicylidenemino)-2-chlorobenzoate]. The compound was further recrystallized repeatedly to get the pure sample. The formation of the compound was confirmed by ^{1}H-NMR and IR spectroscopy and the purity was established by elemental analysis. The details of the experimental procedures along with the spectroscopic data for compound are presented in the experimental part of **Chapter 3**.

5.2.2. Mesomorphic properties:

To investigate the liquid-crystalline behaviour of the synthesised compound, **RN-14-6/Cl**, optical microscopy and differential scanning calorimetric (DSC) studies were carried out. The DSC thermogram, phase transition temperatures and enthalpy associated with the transitions are presented in **Figure 5.10.**

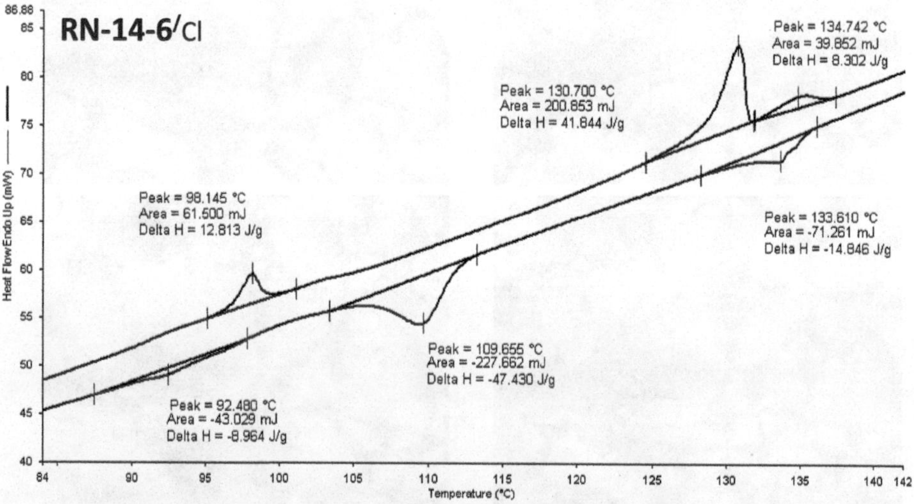

(Heating): K 98.1^{0}C (11.0) K1 130.7^{0}C (35.9) Col_r 134.7^{0}C (7.1)I

(Cooling): K 92.4^{0}C (7.7) K1 109.6^{0}C (40.7) Col_r 133.6^{0}C (12.7)I

Figure 5.10: DSC thermogram, transition temperatures and enthalpies of phase transition of compound **RN-14-6/Cl.**

On slow cooling from the isotropic phase the mesophase grows within the isotropic liquid as telephone-wire structures that are shown in **Figure 5.11a.**

On further cooling, the material enters completely to a structured columnar fan like texture under the polarization microscope. It exhibits pseudo-broken fans or circular domains with extinction brushes directed along polarizing directions. On further cooling the sample, low birefringent stripes appears across the blades of the columnar fan and finally becomes completely low birefringent without changing the columnar texture before it crystallizes. The polarising optical microscopic textures of the compound **RN-14-6/Cl** in a 5.0 μm cell are shown in **Figure 5.11**.

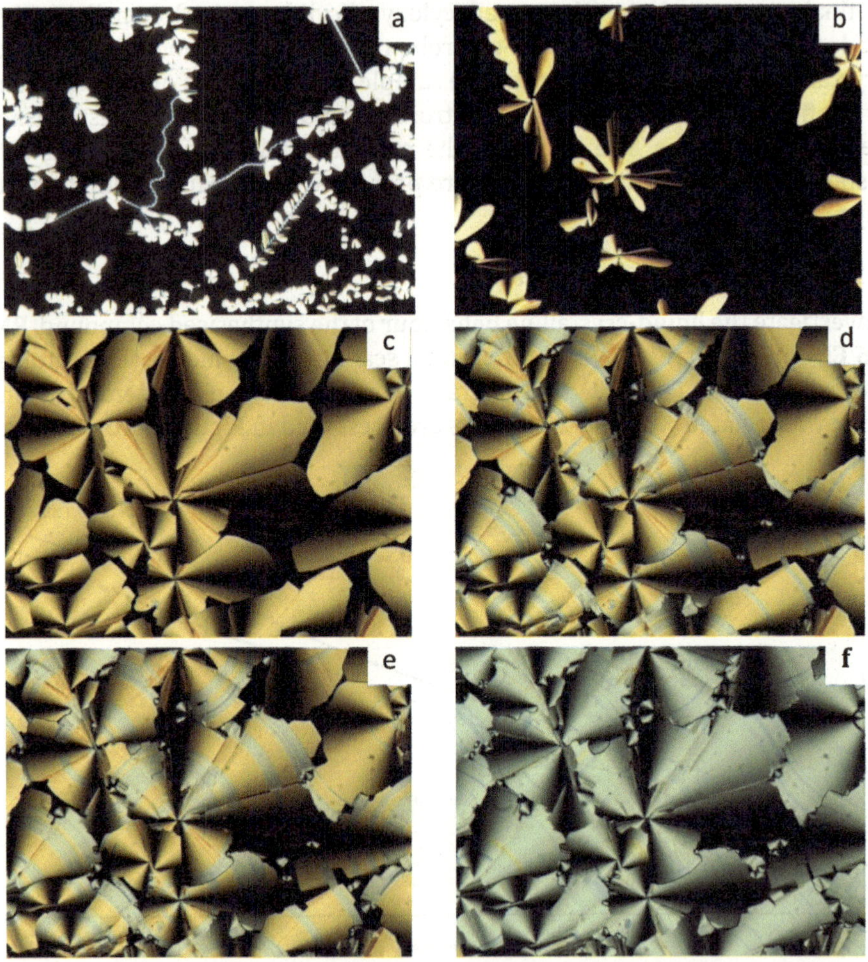

Figure 5.11: Microphotographs of the compound **RN-14-6/Cl** in a 3.5μ cell. Photographs shows (a) helical spiral formation from the isotropic phase to mesomorphic phase at 133.6^{0}C (b) dendritic growth at 133.6^{0}C (c) fully developed columnar texture (d) phase transition appearance of POM texture of the transient transition bars of arcs across the fans at 125.0^{0}C after 3 minutes d) after 5 minutes and e) columnar texture of phase at 133.5^{0}C with a change in colour at 124^{0}C

5.2.3. X-Ray diffraction studies:

However, the growth of the mesophase in telephone-wire structures has been widely related to the existence of 2D periodic structures such as columnar or smectic undulated phases [46-48]. In order to clarify this point, small angle X-ray diffraction experiments were carried out. The measurements were performed at different temperatures within the mesophase on an unaligned sample.

In powder X-ray diffraction measurements the diffuse broad hump observed in the wide angle region ($2\theta \approx 15^0 \sim 25^0$ with a maximum at 4.63Å) from in-plane molecular positional correlations indicating that the layers in this phase is two-dimensional liquid like order (**Figure 5.12**). In contrast several sharp peaks at (1) 70.11 Å, (2) 63.22 Å, (3) 47.46 Å,(4) 40.44 Å, (5) 32.79 Å and (6) 24.41 Å in the small angle region (**Figure 5.13**) are observed in the mesomorphic phase of **RN-14-6/Cl** at 130°C. The first and second peaks in the small angle region have no relation to the first and second order diffraction peaks corresponding to the smectic layer and hence a two dimensional (2D) frustrated layer structure can explain the reflections. Following the hypothesis proposed for 2D rectangular lattices, it is reasonable to assign the most intense peak at 47.46 Å corresponding to layer periodicity and is found to be less than the molecular length of 58.4 Å in most extended conformation shown in **Figure 5.14** although it is likely that somewhat more bent conformation is adopted in the smectic phase. Typical bent core tilt angle of the molecules (tilt angle $\theta \approx 35^0$) and the measured layer spacings would imply an effective molecular length ~58 Å. The reflection in the low angle region (~ $2\theta = 1.26^0$) indicative of in-plane modulation of the smectic layers of wavelength λ ~70.1 Å associated with the 2D polarization splay modulated layer undulated structure, a structural feature associated with the B_7 and the $B1_{RevTilted}$ phases with a 2D lattice of $a \approx$ 70.1 Å (along the layer) and c (layer thickness) ≈ 47.4 Å. The helical filament textures, the multiple peak reflections at low angles, the tilt in molecular layering and non-switchable phase structure are indicative of a $B1_{RevTilted}$ phase structure and suggest that the phase represents 2D polarization splay modulated layer undulated molecular structure of SmCPU family. Further work of electro-optical experiments is in progress to compliment the x-ray results and to assign the reflections and to identify the exact phase structure. The temperature variation of layer thickness is presented in **Figure 5.15** and found only small variation from 47.45 to 47.92 Å.

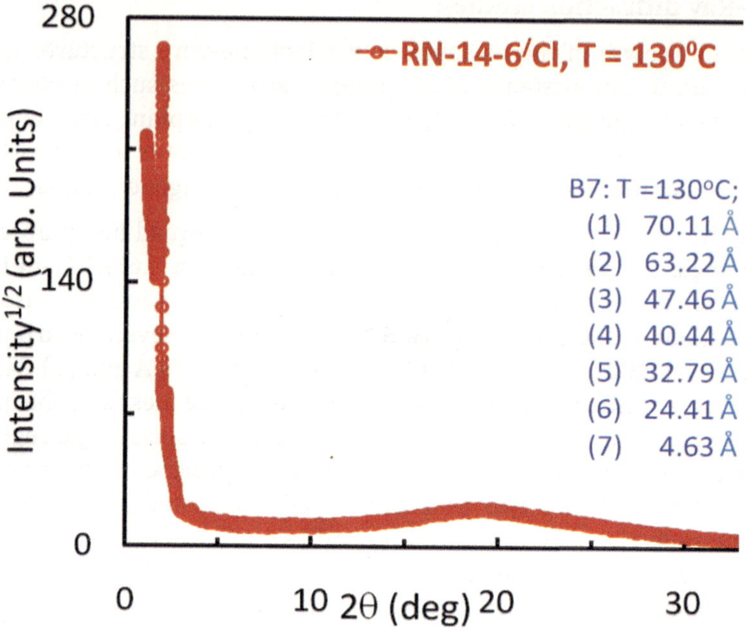

Figure 5.12: X-ray diffraction pattern of **RN-14-6/Cl** at 130 °C

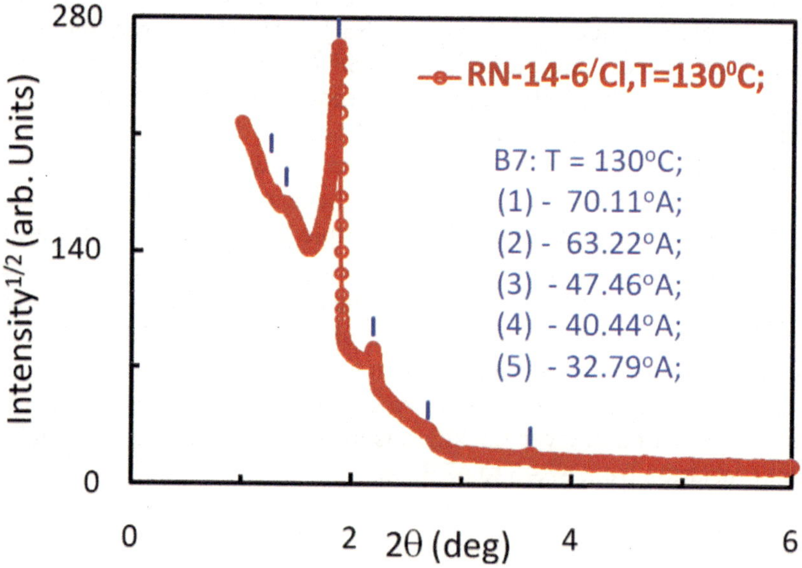

Figure 5.13: X-ray diffraction pattern of **RN-14-6/Cl** at 130 °C in the low angle region

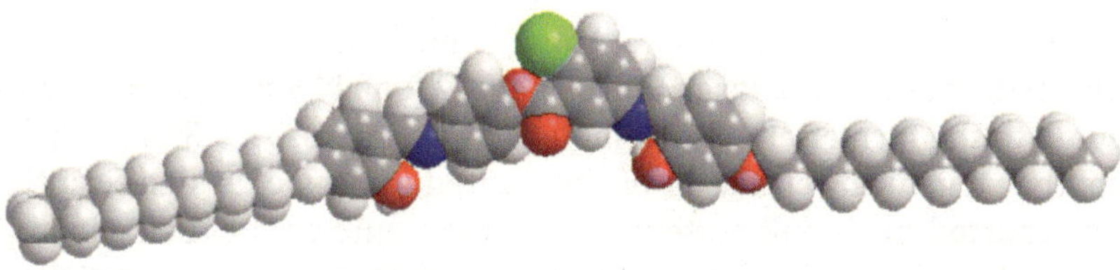

Figure 5.14: Space filling model of energy minimised molecular structure of the molecule **RN-14-6/Cl** with fully extended chain in trans conformation (58.4 Å)

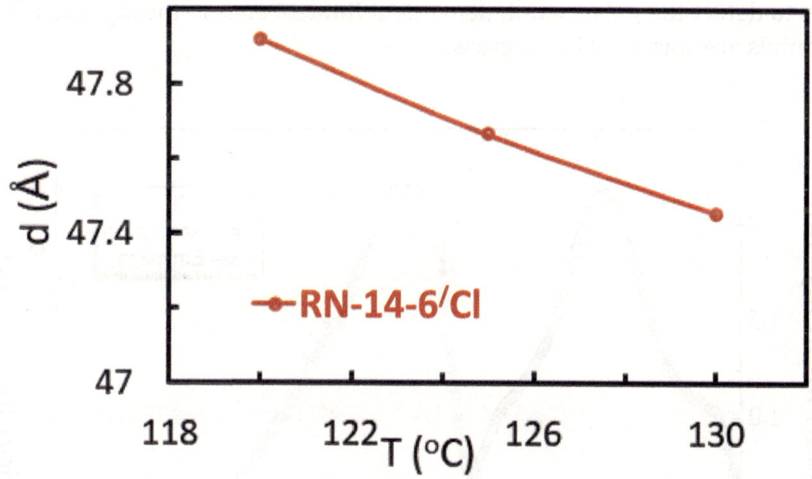

Figure 5.15: Variation of layer thickness as a function of temperature in the LC phase

5.2.4. UV-Visible and fluorescence studies:

The UV absorption and fluorescence spectroscopic properties of compound **RN-14-6/Cl** in solution were studied in various solvents of different concentrations to obtain information regarding absorption and emission maxima and the Stokes shift of fluorescence. The UV-Visible and Fluorescence spectra of a representative compound **RN-14-6/Cl** are shown in **figure 5.16**.The UV-visible spectra in these homologous compounds exhibited strong absorption peaks at ~344 nm with a large molar extinction coefficient (3.60 eV for **RN-14-6/Cl**, ε ~ 216000 Lmol^{-1}cm^{-1},concentration = 10^{-5} M solution in $CHCl_3$) and shoulders at 283 nm, 398 nm and 423 nm. The strong absorption band reflects the π-π* transition of the highly π-conjugated system having the substituted phenyl benzoate unit as the core. Additionally, we investigated the fluorescence spectrum to observe the characteristic features of the excited states of **RN-14-6/Cl**. The fluorescence emission spectrum of **RN-14-6/Cl** was examined in chloroform solution (conc.

1.0 x 10^{-5} M) and it was found to exhibit a strong fluorescence on excitation at 344 nm. The emission maximum (λ_{em}) at 462 nm (2.68 eV) for compound **RN-14-6/Cl** originate from with a large Stokes shift (λ_s) of 115 nm (ΔE = 0.92 eV) was attributed to the formation of intermolecular excimers. This Stokes shift, which reflects the structural relaxation of the excited molecule, is significantly larger than in reported push-pull systems exhibiting liquid crystal behaviour [37-41] indicating that the molecular conformation changes upon excitation. These results are in good agreement with the reported results of molecular J-aggregates [42] in which the excitonic energy is delocalized as a result of intermolecular coupling within the head-to-tail arrangement of the molecules in the solution. The optical band gap investigations of thin films in LC phase and solid state for their possible applications and Fluorescence confocal polarizing microscope studies to detect the polarization dependent fluorescence in mesophase of these compounds are currently in progress.

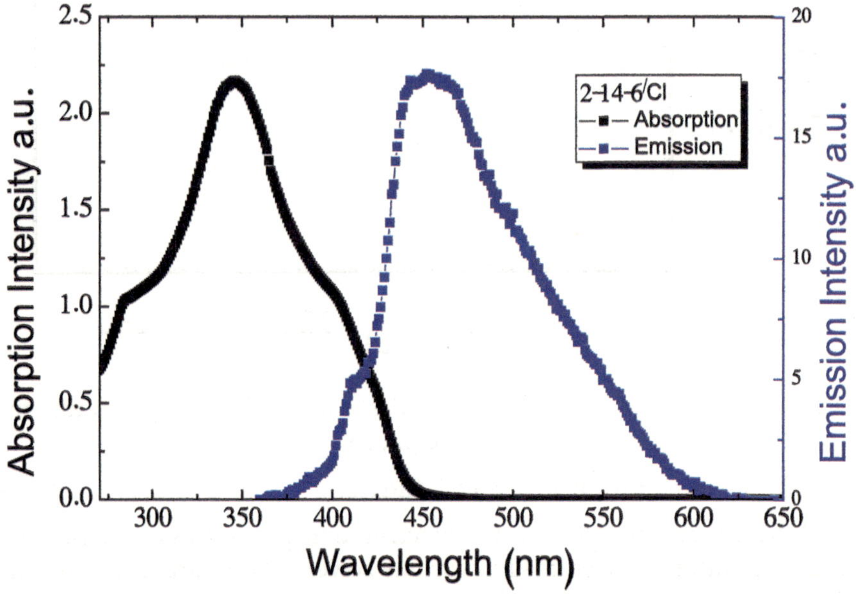

Figure 5.16: UV-Visible and fluorescence spectra of RN-14-6/Cl in chloroform (10^{-5}M)

5.2.5. Synthesis of [4-(N-4-decyloxysalicylidenemino)phenyl]-[3-(N-4-decyloxy salicylideneimine)-4-chlorobenzoate], RN-10-4/Cl :

The starting material in the present study 4-n-decyloxysalicylaldehyde was prepared by Williamson etherification of 2,4-dihydroxybenzaldehyde with n-decyl bromide. The central bent core moiety 4/-aminophenyl-3-amino-4-chlorobenzoate (**5-3c**) was prepared as follows. 4-Chloro-3-nitrobenzoic acid was converted into its acyl chloride followed by condensation with 4-nitrophenol using phase transfer reaction. The resultant 4/-nitrophenyl-4-chloro-3-nitrobenzoate was subjected to 10% Pd-C catalysed reduction to get the desired di-amino bent-core compound. The condensation of 4-n-decyloxy salicylaldehyde with 4-aminophenyl-3-amino-4-chlorobenzoate in presence of a few drops of glacial acetic acid yielded the target bent shaped compounds [4-(N-4/-n-decyloxysalicylidene)-aminophenyl]-[3-(N-4/-n-decyloxysalicylideneamino)-4-chlorobenzoate (**RN-10-4/Cl**). To avoid the formation of side products, the precipitated compounds were filtered when the solution was hot to yield the pure compounds. The compounds were further recrystallized repeatedly to get the pure samples. The formation of all of the compounds was confirmed by ^{1}H-NMR and IR spectroscopy and the purity was established by elemental analysis. The other homologues with n = 12, 14 and 16 were also synthesized following the same procedure and characterized. The details of the experimental procedures along with the spectroscopic data for all the homologous series of compounds are presented in the experimental part of **Chapter 3**. The liquid-crystalline behaviour of the synthesised compounds had been investigated by optical microscopy and differential scanning calorimetry (DSC).

5.2.6. Mesomorphic properties:

All the compounds of the homologous series (**RN-n-4/Cl**) show mesomorphic behaviour for a moderate temperature range. The lower homologues of the series (**RN-10-4/Cl** and **RN-12-4/Cl**) show B1 mesophases whereas higher homologue (**RN-14-4/Cl** and **RN-16-4/Cl**) shows B7 phases. The differential scanning thermograms of all the compounds of the homologous series show enantiotropic phase transition. The transition temperatures, enthalpies and entropies associated with the phase transitions of all the compounds of the homologous series **RN-n-4/Cl** (n = 10, 12, 14 and 16) obtained from DSC at a scan rate of 5^{0}C min^{-1} in the second heating and first cooling scans are presented in **Table 5.2.** Mesomorphic range for the series of compounds in heating and cooling scan are presented in **Figure 5.17**. All the compounds are enantiotropic with a short mesophase range in heating and long range in cooling. DSC thermogram of a representative compound **RN-14-4/Cl** shows two transitions in heating cycle at 131.33 (melting point, ΔH = 47.78 kJ/mol, ΔS = 118.17J/K/mol) and 137.45 (clearing point, ΔH = 17.04 kJ/mol, ΔS = 41.51J/K/mol) and two transition in cooling cycle at 136.73 (ΔH = 16.53 kJ/mol, ΔS = 40.34J/K/mol) and 111.87 (ΔH = 42.85 kJ/mol, ΔS = 111.33J/K/mol). Other transitions at 99.48, 104.6 in heating cycle and 94.94 and 91.71 in cooling cycles are crystal to crystal transitions. (**Figure 5.18**).

Table 5.2: Transition Temperatures (⁰C), Enthalpies (*Italic*, kJ/mol) and entropies (*Italic*, J/mol/K) of homologous series **RN-n-4'Cl**. Compound RN-10-4'Cl and RN-12-4'Cl shows B1 mesophase whereas compound RN-14-4'Cl and RN-16-4'Cl shows B7 mesophase.

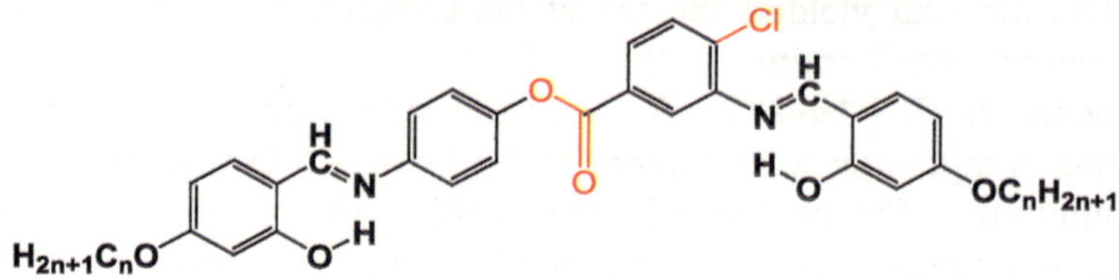

Compound	R	Cr	Heating Cooling	B1/ B7	Heating Cooling	I	ΔT
2-10-4'Cl*	$C_{10}H_{21}$	•	127.76 [*20.2, 50.4*]	•	129.53 -	•	1.77
		•	102.23 -	•	127.31 [*8.5, 21.2*]	•	25.09
2-12-4'Cl*	$C_{12}H_{25}$	•	133.28 [*37.1, 91.4*]	•	134.88 [*0.2, 0.5*]	•	1.60
		•	110.61 [*34.7, 90.3*]	•	132.98 [*6.3, 15.6*]	•	22.37
2-14-4'Cl	$C_{14}H_{29}$	•	131.33 [*47.8, 118.2*]	•	137.45 [*17.0, 41.5*]	•	6.12
		•	111.87 [*42.8, 111.3*]	•	136.73 [*16.5, 40.3*]	•	24.86
2-16-4'Cl	$C_{16}H_{33}$	•	130.03 [*41.2, 102.2*]	•	132.40 [*1.4, 3.5*]	•	2.37
		•	108.73 [*54.2, 141.9*]	•	131.34 [*6.9, 17.2*]	•	22.61

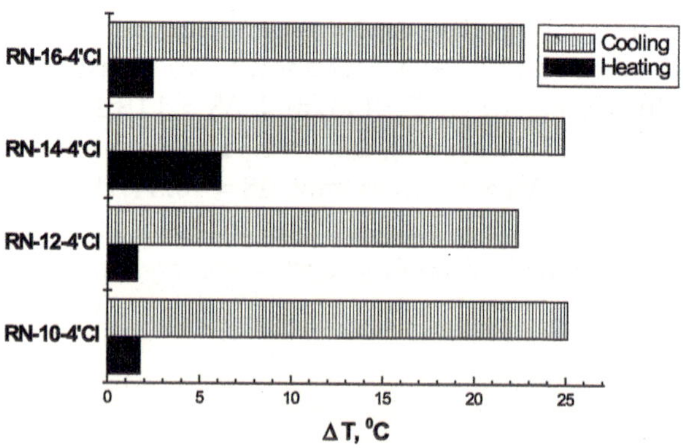

Figure 5.17: Mesomorphic range (ΔT, ⁰C) for the series of compounds in heating and cooling scan.

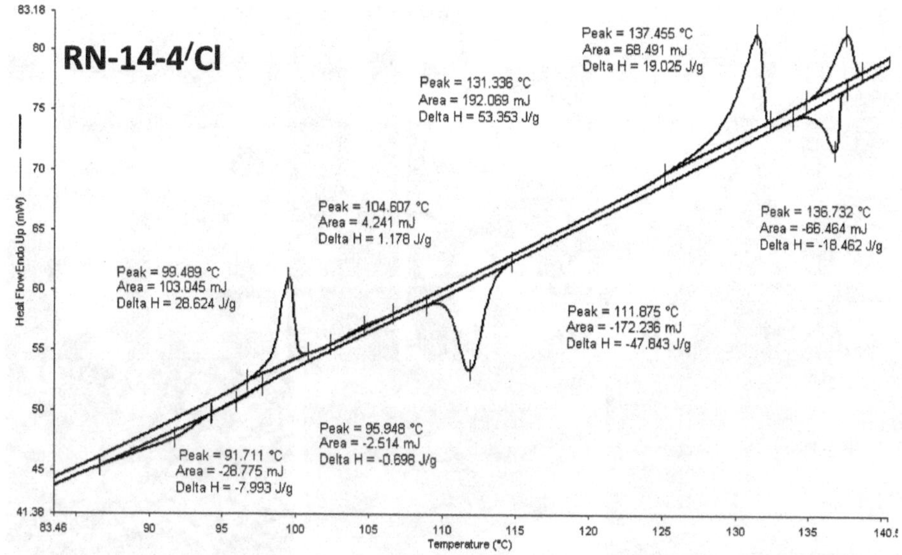

Figure 5.18: DSC thermogram of **RN-14-4/Cl** on second heating and cooling at a scan rate 5 ^{0}C min^{-1}

Polarising optical microphotographs of the compounds of the homologous series are presented in **Figure 5.19**. Upon very slow cooling the isotropic liquid, the mesophase defect textures exhibited by **RN-16-4/Cl (Figure 5.19a-f)** resembles the B7 phase [33,47-58] texture of fascinating long spiral domains. The microscopic texture of the B7 phase is extraordinary and not comparable with the textures of other B7 phases. If the isotropic liquid is cooled down very fast, the B7 phase appears as a cluster of thin thread-like or straight lancet-like nuclei which coalesce to an unspecific texture. On very slow cooling, variety of texture variants is obtained, but these mostly occur simultaneously within the same preparation. Also, on slow cooling of the isotropic liquid, the B7 phase frequently forms elongated germs such as straight lancet-like or irregular thread-like germs. But frequently these germs have clearly a spiral or double-spiral character. **Figure 5.19g** shows the growth of a spiral germ. The handedness of the screw-like nuclei could be determined by adjusting the focus of the microscope. The B7 phase frequently grows as oval or sometimes as circular domains. Such domain (shown in **figure 5.19a, b, f**)exhibits equidistant stripes.

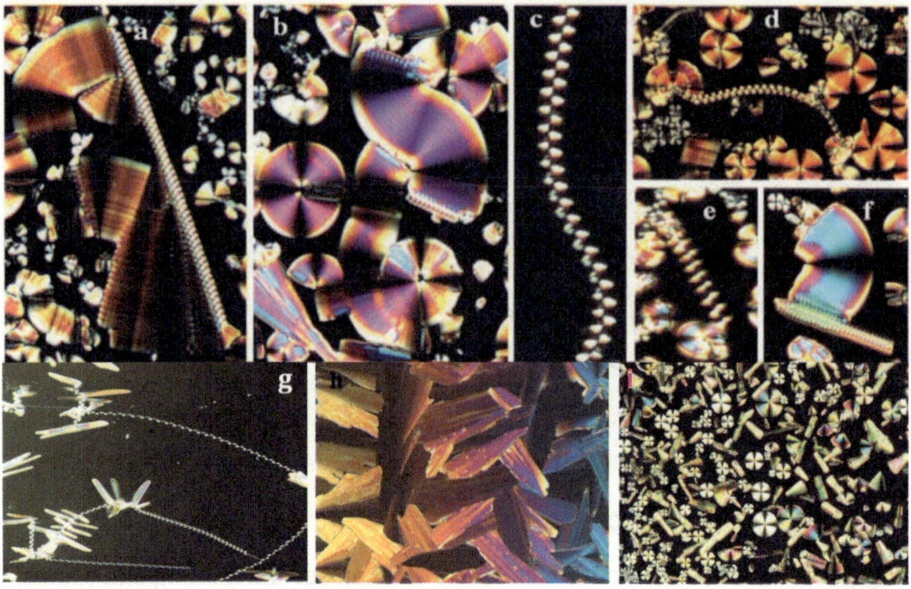

Figure 5.19: Polarising optical microphotographs of various features of B7 textures of compounds **RN-16-4/Cl** sandwiched between glass plate and cover slip (a-g). (h) **RN-16-4/Cl** in a 5μm cell at 132⁰C. (i) texture of **RN-10-4/Cl** at 134.2⁰C.

In nylon 6,6 treated cell of the compound **RN-14-4/Cl,** straight lancet like or spiral germs appears when slowly cooled the sample below isotropic temperature. On further cooling the germs coalesce to banana leaf, ribbons and fanlike textures with equidistant stripes. HTAB treated film appears as blue fan on slow cooling the isotropic liquid. On further cooling, pink stripes appears across the blades of the fan and finally becomes completely pink in colour. In some part of the cell, B7 texture appears as weakly birefringent fan shaped texturewhich reminds one of the corresponding texture of a SmA phase (**Figure 5.20d**). Sometimes, but notalways, within the fans equidistant black stripes arevisible between crossed polarizers. [47]

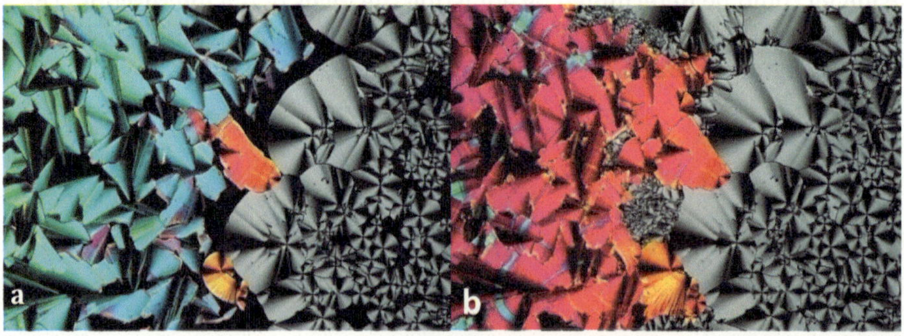

Figure 5.20: Polarising optical photographs of **RN-14-4/Cl** on a HTAB treated cell at a)133.2 ^{0}C, b) 124.8 ^{0}C, C) Nylon 6,6 treated cell at 131 ^{0}C, d) Different side of HTAB treated cell at 124.8 ^{0}C.

5.2.7. X-Ray Diffraction study:

RN-10-4/Cl

To confirm the mesophase structures X-ray diffraction studies are performed on a representative compound belonging to the homologous series **RN-*n*-4/Cl.** A typical diffraction pattern obtained for the compound **RN-10-4/Cl** at 120^{0}C is shown in **Figure 5.21.** The diffraction pattern indicated a strong reflection at 2θ = 1.95^0 corresponding to 45.2Å. This corresponds to the molecular length n which is slightly lower than the estimated molecular length of 48.7 Å (**Figure 5.22**) in all trans conformation indicated a tilted layer with a tilt angle 16^0. In the small angle region (2θ = 1 - 8^0) there are other low intense peaks as observed for the 2D polarization splay modulated layer undulated structure, a structural feature associated with the B_7 and the $B1_{RevTilted}$ phases with a 2D lattice. In addition to these reflections a broad diffuse wide angle reflection at 2θ = 19.21^0 corresponding to 4.51 Å reflects the spacing between the aliphatic chains indicative of a liquid-like in-plane order of the fluid phase. The formation of stripes on focal conic textures, the multiple peak reflections at low angles, the tilt in molecular layering and non-switchable phase structure are indicative of a $B1_{RevTilted}$ phase structure and suggest that the phase represents 2D polarization splay modulated layer undulated molecular structure of SmCPU family. Further work of electro-optical experiments is in progress to compliment the x-ray results and to assign the reflections and to identify the exact phase structure.

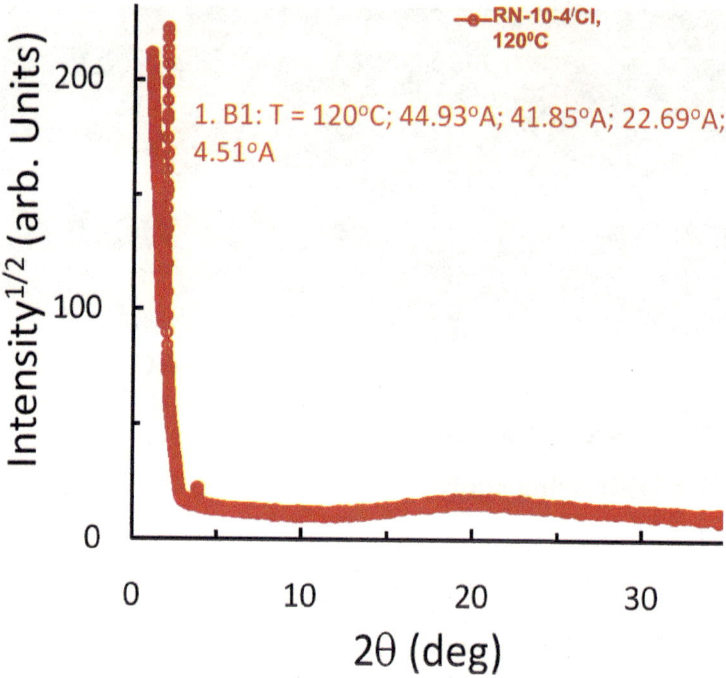

Figure 5.21: X-ray diffraction pattern of **RN-10-4/Cl.**

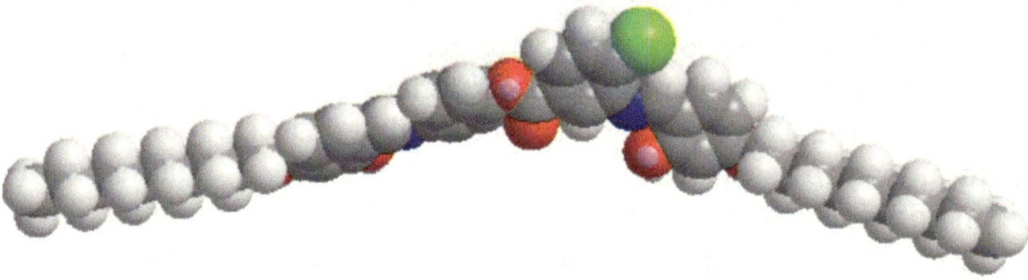

Figure 5.22: Fully extended all trans confotmation of **RN-10-4/Cl.**(L= 46.72)

5.2.8. UV-Visible and Fluorescence study:

The UV absorption and fluorescence spectroscopic properties of compound **RN-10-4/Cl** in solution were studied in various solvents of different concentrations to obtain information regarding absorption and emission maxima and the Stokes shift of fluorescence. The UV-Visible and Fluorescence spectra of a representative compound **RN-10-4/Cl** are shown in **Figure5.23**.The UV-visible spectra in these homologous compounds exhibited strong absorption peaks at ~344 nm with a large molar extinction coefficient (3.60 eV for **RN-10-4/Cl**, ε ~ 98300 $Lmol^{-1}cm^{-1}$,concentration = 1 X 10^{-5} M solution in $CHCl_3$) and shoulders at 283 nm, 398 nm and 423 nm. The strong absorption band reflects the π-π* transition of the highly π-conjugated system having the substituted phenyl benzoate unit as the core. Additionally, we investigated the fluorescence spectrum to observe the characteristic features of the excited states of **RN-10-4/Cl**. The fluorescence emission spectrum of **RN-10-4/Cl** was examined in chloroform solution (conc. 1.0 x 10^{-5} M) and it was found to exhibit a strong fluorescence on excitation at 344 nm. The emission maximum (λ_{em}) at 450 nm (2.7 eV) for compound **RN-10-4/Cl** originate from with a large Stokes shift (λ_s) of 106 nm (ΔE = 0.84 eV) was attributed to the formation of intermolecular excimers. This Stokes shift, which reflects the structural relaxation of the excited molecule, is significantly larger than in reported push-pull systems exhibiting liquid crystal behaviour [37-41] indicating that the molecular conformation changes upon excitation. These results are in good agreement with the reported results of molecular J-aggregates [42] in which the excitonic energy is delocalized as a result of intermolecular coupling within the head-to-tail arrangement of the molecules in the solution. The optical band gap investigations of thin films in LC phase and solid state for their possible applications and Fluorescence confocal polarizing microscope studies to detect the polarization dependent fluorescence in mesophase of these compounds are currently in progress.

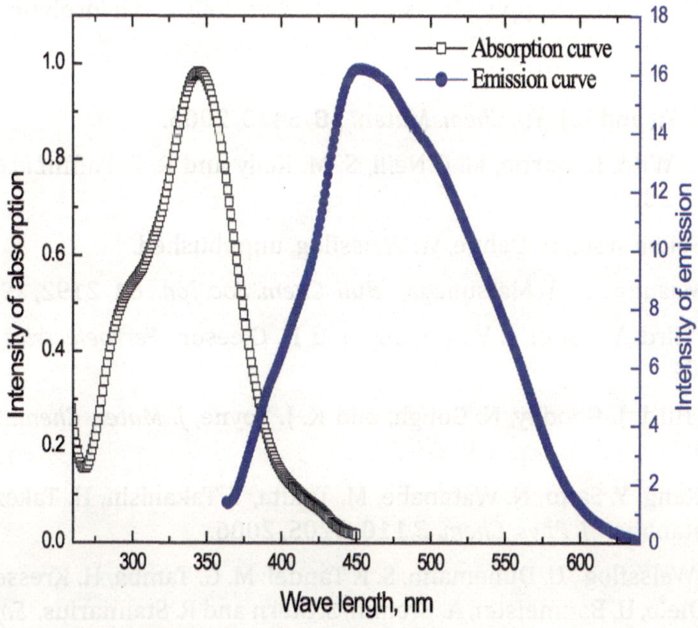

Figure 5.23: UV-Vis and Fluorescence curve of 2-10-4/Cl in chloroform solution (10^{-5} M).

5.3. Conclusion:

The influence of polar (fluoro and chloro) substitution at the bent aromatic core was investigated with respect to its effects upon the structures of the mesophases formed by 4-ring bent-core molecules. It was found that, formation of different mesophases is not solely determined by the types of substituent but also by their position. Substitution by chloro and fluoro group on the four ring system produced liquid crystals with a wide variety of mesophases.

Combining the XRD data with DSC and POM results provide unequivocal identification of the mesophase of RN-n-6/F series as B1 columnar phase. Chloro substituted homologues series of compounds, RN-n-6/Cl (n=14) and RN-n-4/Cl (n = 10, 12, 14, 16) belong to the class of compounds exhibiting polarisation modulated layer undulated (PMLU) phase and show $B1_{Rev.Tilted}$ and B7 phase like textures.

Because of the possible free rotation of the central ester linkage, all the rings areat different plane with respect to each other. The twisted conformation of the molecule and the hydroxyl groups in the side wings of these molecule participate in inter and or intra molecular H-bonding which in turn give rise to the formation of this kind of undulated and modulated smectic phases. Additionally, all the materials described in this chapter have interesting intrinsic light generating capacity in the region 450-580 nm still retaining their banana polymorphism which could be of use in applications; they are thermally and hydrolytically stable.

References:

1. F. C. Yu and L. J. Yu, *Chem. Mater.*, **18**, 5410, 2006.
2. J. H. Wild, K. Bartle, M. O'Neill, S. M. Kelly and R. P. Tuffin*Liq.Cryst.*, **33**, 635, 2006.
3. S. Sokolowski, H. Dehne, W. Weissflog, unpublished.
4. T. Masuda and Y. Matsunaga, *Bull. Chem. Soc. Jpn.*, **64**, 2192, 1991.
5. M. Hird, Y. Raoul, J. W. Goodby and H. Gleeson, *Ferroelectrics,* **309**, 95, 2004.
6. M. Hird, J. Goodby, N. Gough, and K. J. Toyne, *J. Mater. Chem.* **11**, 2732, 2001
7. S. Kang, Y. Saito, N. Watanabe, M. Tokita, Y. Takanishi, H. Takezoe and J. Watanabe, *J. Phys. Chem. B,***110**, 5205, 2006.
8. W. Weissflog , U. Dunemann, S. F. Tandel, M. G. Tamba, H. Kresse, G. Pelzl, S. Diele, U. Baumeister, A. Eremin, S. Stern and R. Stannarius, *Soft Mattar,* **5**, 1840, 2009.
9. U. Dunemann, M. W. Schroder, R. A. Reddy, G. Pelzl, S. Diele, W. Weissflog, *J. Mater. Chem.,***15**, 4051, 2005.
10. G. Dantlgraber, D. Shen, S. Diele, C. Tschierske, *Chem. Mater.,* **14**, 1149, 2002.
11. D. Shen, A. Pegenau, S. Diele, I. Wirth, C. Tschierske,*J. Am. Chem. Soc.,* **122**, 1593, 2000.

12. D. Shen, S. Diele, G .Pelz, I. Wirth, C. Tschierske, *J. Mater. Chem.*, **9**, 661, 1999.

13. D. Shen, S. Diele, I. Wirth, I. Tschierske, *Chem. Commun.*, 2573, 1998.

14. W. Weissflog, L. Kovalenko, I. Wirth, S. Diele, G. Pelzl, H. Schmalfuss, H. Kresse, *Liq.Cryst.*, **27**, 677, 2000.

15. H. T. Nguyen, J. C. Rouillon, J. P. Marcerou, J. P. Bedel, P. Barois, S. Sarmento, *Mol. Cryst. Liq. Cryst.*, **328**, 177, 1999.

16. W. Weissflog, H. Nadasi, U. Dunemann, U. Pelzl, S. Diele, A. Eremin, H. Kresse, *J. Mater. Chem.*, **11**, 2748, 2001.

17. G. Pelzl, S. Diele, S. Grande, A. Jakli, C. Lischka, H. Kresse, H. Schmalfuss, I. Wirth, W. Weissflog, *Liq. Cryst.*, **26**, 401, 1999.

18. S. Shubashree, B. K. Sadashiva, S. Dhara, Liq. Cryst., **29**, 789, 2002.

19. I. Wirth, S. Diele, A. Eremin, G. Pelzl, L. Kovalenko, N. Pancenko, S. Grande, W. Weissflog, *J. Mater. Chem.*, **11,** 1642, 2001.

20. W. Weissflog, Ch. Lischka, I. Benne, T. Scharf, G. Pelzl, S. Diele, H. Kruth, *Proc. SPIE*, **14**, 3319, 1998.

21. G. Pelzl, M.W. Schroder, U. Dunemann, S .Diele, W. Weissflog, C. Jones, D. Coleman, N.A. Clark, R. Stannarius, J. Li, B. Das, S. Grande, *J. Mater. Chem.*, **14**, 2492, 2004.

22. H. Nadasi, W. Weissflog, A. Eremin, G. Pelzl, S. Diele, B. Das, S. Grande, *J. Mater. Chem.*,**12**, 1316, 2002.

23. R. A. Reddy, B. K. Sadashiva, *J. Mater. Chem.*, **14**, 1936, 2004.

24. H. N. S. Murthy, B. K. Sadashiva, *J. Mater. Chem.*, **14**, 2813, 2004.

25. C. V. Yelamaggad, I. S. Shashikala, U. S. Hiremath, G. Liao, A. Jakli, D. S. S. Rao, S. K. Prasad, Q. Li, *Soft Matter*, **2**, 785, 2006.

26. B. E. Smart, *In Organofluorine Chemistry Principles and CommercialApplications*, R. E. Banks, B. E. Smart, and J.C. Tatlow, Eds.; Plenum Press: New York, (1994), p. **57**.

27. F. Guittard, S. J. Geribaldi, *Fluorine Chem.*, **107**, 363, 2001.

28. F. Guittard, E.T. deGivenchy, S. Geribaldi, A. J. Cambon, *Fluorine Chem.*, **100**, 85, 1999.

29. D. Kardas, M. Prehm, U. Baumeister, D. Pociecha, R. A. Reddy, G. H. Mehl and Ch. Tschierske, *J. Mater. Chem.*, **15**, 1722, 2005.

30. H. N. S. Murthy and B. K. Sadashiva, *J. Mater. Chem.*, **15**, 2056, 2005.

31. R. A. Reddy, V. A. Raghunathan and B. K. Sadashiva, *Chem. Mater.*, **17**, 274, 2005.

32. J. P. Bedel, J. C. Rouillon, J. P. Marcerou, M. Laguerre, H. T. Nguyen and M. F. Achard, *Liq.Cryst.*, **27,** 1411, 2000.

33. J. P. Bedel, J. C. Rouillon, J. P. Marcerou, M. Laguerre, H. T. Nguyen and M. F. Achard, *J. Mater. Chem.***12**, 2214, 2002.

34. G. Pelz, W. Weissflog, U. Baumeister and S. Diele, *Liq. Cryst.*, **30**, 1151, 2003.

35. R. A. Reddy and B. K. Sadashiva, *Liq.Cryst.*, **30**, 1031, 2003.

36. H. Takezoe and Y. Takanishi, *Jap. J. Appl. Phys.*,**45**, 597, 2006.

37. H. Gallardo, R. Cristiano, A. A. Vieira, R. A. W. Neves Filho, R. M. Srivastava and I. H. Bechtold, *Liq.Cryst.*, **35**, 57–63, 2008.

38. A. A. Vieira, R. Cristiano, A. J. Bortoluzzi and H. Gallardo, *J. Mol. Struct.*, **875**, 364–371, 2008.

39. R. M. Srivastava, R. A. W. Neves Filho, R. Schneider, A. A. Vieira and H. Gallardo, *Liq. Cryst.*, **35**, 737–742, 2008.

40. R. Cristiano, F. Ely and H. Gallardo, *Liq. Cryst.*, **32**, 15–25, 2005.

41. A. A. Vieira, R. Cristiano, F. Ely and H. Gallardo, *Liq.Cryst.*, **33**, 381–390, 2006.

42. F. Camerel, L. Bonardi, M. Schmutz and R. Ziessel, *J. Am. Chem. Soc.*, **128**, 4548–4549, 2006.

43. W. Weissflog, U. Dunemann, M. W. Schroder, S. Diele, G. Pelzl, H. Kresse and S. Grande, *J. Mater. Chem.*, **15**, 939, 2005.

44. W. Weissflog, S. Sokolowski, H. Dehne, B. Das, S. Grande, M. W. Schröder, A. Eremin, S. Diele, G. Pelzl and H. Kresse, *Liq.Cryst.*, **31**, 923, 2004.

45. S. Kang, J. Thisayukta, H. Takezoe, J. Watanabe, K. Ogino, T. Doi and T. Takahashi, *Liq.Cryst.*, **31**, 1323, 2004.

46. D. A. Coleman, J. Fernsler, N. Chattham, M. Nakata, Y. Takanishi, E. Körblova, D. R. Link, R. F. Shao, W. G. Jang, J. E. Maclennan, O. Mondainn-Mova, C. Boyer, W. Weissflog, G. Pelzl, L. C. Chien, J. Zasadzinski, J. Watanabe, D. M. Walba, H. Takezoe, N. A. Clark, *Science*, **301**, 1204, 2003.

47. G. Pelzl, S. Diele, A. Jakli, Ch. Lischka, I. Wirth, W. Weissflog, *Liq. Cryst.*, **26**, 135, 1999.

48. H. N. Shreenivasa Murthy, B. K. Sadashiva, *Liq.Cryst.*, **30**, 1051, 2003.

49. C. K. Lee and L. C. Chien, *Liq.Cryst.*, **26**, 609, 1999.

50. G. Heppke, D. D. Parghi and H. Sawade, *Liq. Cryst.*, **27**, 313-320, 2000.

51. A. Jakli, Ch. Lischka, W. Weissflog, G. Pelzl and A. Saupe, *Liq. Cryst.*, **27**, 1405, 2000.

52. C. K. Lee, A. Primak, A. Jakli, E. J. Choi, W. C. Zin and L. C. Chien, *Liq.Cryst.*, **28**, 1293, 2001.

53. J. Mieczkowski, J. Szydlowska, J. Matraszek, D. Pociecha, E.Gorecka, B. Donnio and D. Guillon, *J. Mater. Chem.*, **12**, 3392, 2002.

54. A. Jakli, D. Kruerke and G. G. Nair, *Phys. Rev. E*, **67**, 051702, 2003.

55. R. A. Reddy and B. K. Sadashiva, *Liq. Cryst.*, **30**, 273, 2003.

56. Y. A. Nastishin, M. F. Achard, H. T. Nguyen and M. Kleman, *Eur Phys. J. E*, **12**, 581, 2003.

57. M. F. Achard, M. Leman, Y. A. Nastishin and H. T. Nguyen, *Eur. Phys. J. E*, **16**, 37, 2005.

58. C. V. Yelamaggad, M. Mathews, S. A. Nagamani, D. S. S. Rao, S. K. Prasad, S. Findeisen and W. Weissflog, *J. Mater. Chem.*.**17**, 284-298, 2007.

CHAPTER 6

Effect of Non-Polar (Methyl) Substituent on Mesomorphism

In this chapter, the effect of methyl (non-polar) substituents at various position of the central bent core unit of achiral four ring banana shaped molecules are discussed. The phases exhibited by these compounds were confirmed by POM, DSC and XRD techniques.

6. Effect of non-polar substituent on four ring mesogens:

The nature of mesomorphic behaviour and molecular structural arrangement can be influenced by the introduction of lateral substituents in different positions of the phenyl rings of the molecules exhibiting calamitic and discotic liquid crystal behaviour. Depression of melting point as well as enhancement of mesomorphic range can be achieved by the introduction of lateral substituent in such molecules. The introduction of lateral substituents in the rod-like molecule not only decreases the length-to-breath ratio but also the clearing temperatures. Hence the laterally substituted rod like molecules with a change in molecular shape manifests distinct packing of the molecules and can lead to new liquid crystalline phases in addition to the existing phases depending on the interactions between molecules. [1] Banana-shaped mesogens can be laterally substituted by atoms or small groups only in a limited way, depending on the number of aromatic rings. One has to distinguish between the substitution at the central phenylene ring and that at the outer rings, and in both cases the position of lateral substituents is of great influence. These two substitutions are basically different, because the substituents near the linking groups of the central phenyl ring can change the bending angle between the two legs of the molecules. Systematic studies performed in the parent series of five-ring banana-shaped mesogens prove the sensitivity of such compounds to lateral substituents attached to the central ring. Groups like cyano, methyl or methoxy at the top of the bent molecules or between the two legs of the molecule that is parallel or antiparallel to the bent direction of the molecules, prevent the existence of any mesophases. [2]

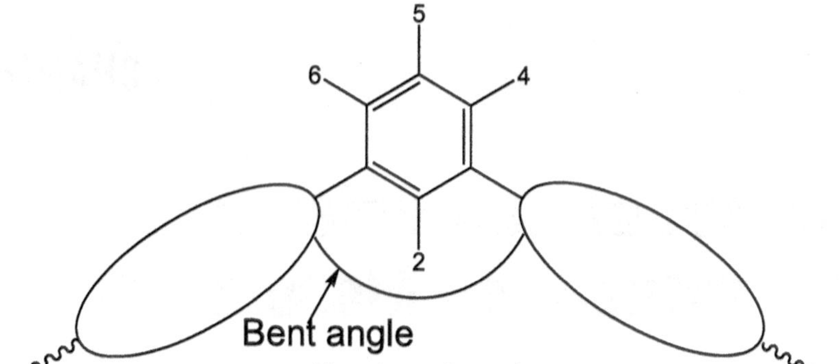

Bent molecule

The introduction of small functional groups like methyl or nitro in 2-position between the two arms of the molecule i.e. into the acute angle of the molecules, results in a high mesophase stability. The 2-methyl substituted derivatives exhibited

B5 phase. [3] The 2-nitro substituted homologues exhibit the B7 phase up to about 175 ^{0}C. [4] The clearing temperatures of these 2-substituted compounds are comparable with those of the unsubstituted parent compounds. [5] 4-substituted chloro compounds exhibited a large thermal range of B2 phase ~70-120^{0}C. [6] In contrast 4,6-dichloro substitution in these compounds revealed unexpected phase structures related to calamitic like nematic and smectic phases replacing banana phases. [7] Examination of these compounds by NMR investigations revealed a change in conformation and an increase in the bending angle from 120^0 in parent compounds to 134^0 and 165^0 in mono and disubstituted compounds respectively. The nearly stretched conformation reported in the crystalline state [8] and the large bending angle in disubstituted compounds justified the observation of calamitic phases. However large alkyl chain substituent viz. n-hexyl moiety in 4-position [9] revealed nematic phases in these compounds on supercooling. In compounds with polar 4-cyano- substituent novel phase variants of SmA–SmC–B2 and N–SmA–SmC–B2 were found. [10-11] The substitution in the 5-position by a methoxy-carbonyl moiety promoted LC behaviour provided the molecule possesses perfluoroalkyl groups in the terminal positions. [12]

Hence the influence of the substituents attached to the central aromatic unit makes it possible to vary the bending angle in a delicate way and hence modify the possible interactions between the aromatic cores and aliphatic chain interactions to manifest unexpected novel mesophases.

In this chapter the influence of a non-polar substituent like methyl group in 2- and 4- positions in unsymmetrical four-ring bent-core compounds are studied and the results are presented.

The newly designed bent core and its derivatives are presented in **Table 6.1**

<table>
<tr><th colspan="2">Table 6.1: Modified Structures of central bent core unit with a substituent</th></tr>
<tr><td>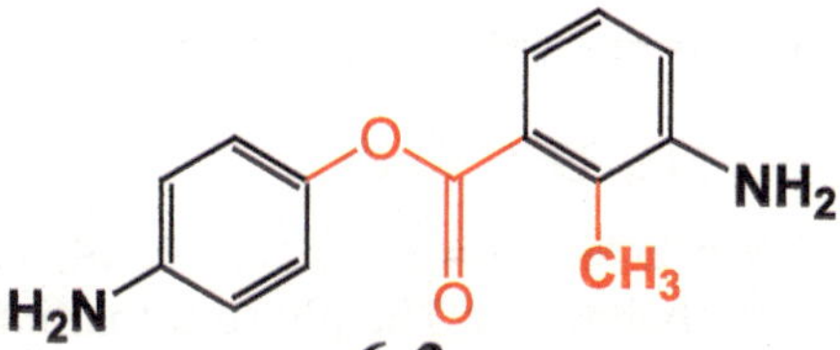
6-3a</td><td>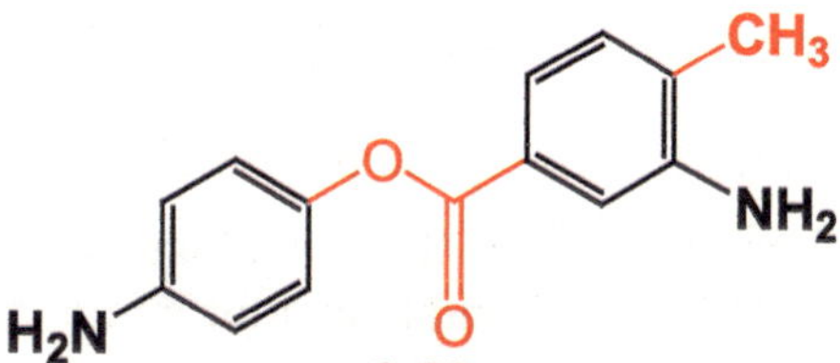
6-3b</td></tr>
<tr><td>4/-aminophenyl-4-Chloro-3-aminobenzoate</td><td>4/-aminophenyl-2-Methyl-3-aminobenzoate</td></tr>
<tr><td colspan="2">Designed compounds</td></tr>
<tr><td colspan="2">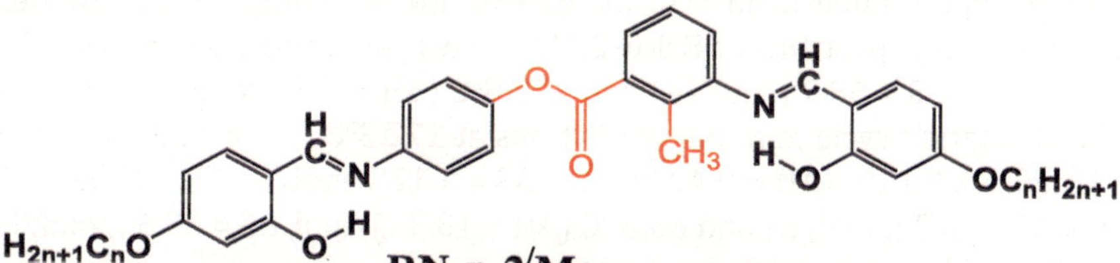
RN-n-2/Me: n = 6, 8, 10, 12, 14, 16
[4-(N-4/-n-alkyloxysalicylidene)-aminophenyl]-[2-methyl-3-(N-4/-n-alkyloxysalicylideneamino)-benzoateRN-n-2/Me</td></tr>
<tr><td colspan="2">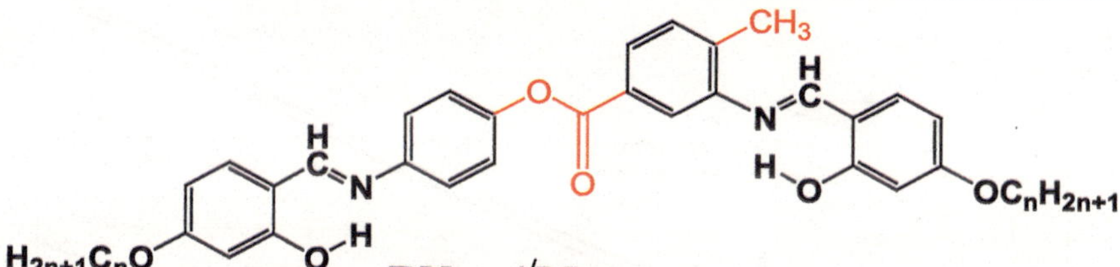
RN-n-4/Me: n = 14
[4-(N-4/-n-alkyloxysalicylidene)-aminophenyl]-[4-methyl-3-(N-4/-n-alkyloxysalicylideneamino)-benzoateRN-n-4/Me</td></tr>
</table>

6.1. [4-(N-4-alkyloxysalicylidenemino)phenyl]-[3-(N-4-alkyloxysalicylidene- amino)-2-methylbenzoate], RN-n-2/Me

6.1.1. Synthesis:

The synthesis of this compound is carried out by extending the two legs of the diamine central core by decyloxy salicylaldehyde units. 4-methyl-3-nitrobenzoic acid is converted to its acyl chloride by the reaction of thionyl chloride. This acyl chloride is condenced with 4-nitrophenol to get dinitro product which on reduction gives the diamine central core (**6a**). The side wing, decyloxy salicylaldehyde is synthesised by Williamson etherification of 2,4-dihydroxybenzaldehyde with n-decyl bromide. Schiff's base condensation of the diamine core and decyloxy salicylaldehyde is carried out in absolte ethanol in presence of few drops of acetic acid as catalyst. The product is filtered off in warm condition and recrystallised several times. The detailed synthetic procedure is presented in **Chapter 3.**

6.1.2. Mesomorphic properties of RN-6-2/Me, DSC study:

All compounds of the homologous series **RN-n-2/Me** (n = 6, 8, 10, 12, 14, 16 and 18) were found to exhibit liquid crystalline behaviour. The phase transition temperatures, enthalpy and entropy associated with the phase transitions of **RN-n-2/Me** are presented in **Table 6.2.** The compound with shorter alkoxy chain **RN-n-2/Me** (n = 6 and 8) shows both nematic and smectic mesophases whereas compounds with higher chain length (n=10-18) shows only smectic mesophases. The transition temperatures, enthalpies and entropies associated with the phase transitions as obtained from DSC for **RN-6-2/Me** at 5^0 min^{-1} are presented in **Figure 6.1.** DSC spectrum of **RN-6-2/Me** shows two transitions at 95.2°C (ΔH = 21.7 kJ/mol, ΔS = 58.9 J/mol/K) and 172.8°C (ΔH = 0.30. kJ/mol, ΔS = 0.68 J/mol/K) in heating cycle and four transitions at 172.3°C (ΔH = 0.33 kJ/mol, ΔS = 0.74 J/mol/K), 90.1°C (ΔH = 0.40 kJ/mol, ΔS = 1.12 J/mol/K), 88.6°C (ΔH = 0.99 kJ/mol, ΔS = 2.72 J/mol/K) and 65.6 °C (ΔH = 13.3 kJ/mol, ΔS = 39.5 J/mol/K) in the cooling cycle.

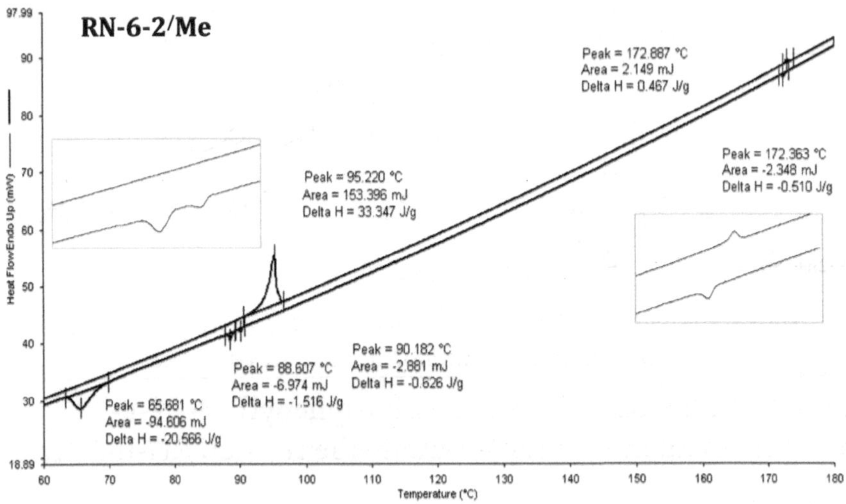

Figure 6.1: DSC thermogram of RN-6-2/Me. Inset shows enlarged image of selected portion of the graph.

Table 6.2: Transition temperatures obtained from differential scanning calorimetry at a scan rate of 5⁰C/min in 2nd heating and cooling data for **RN-6-2/Me**

Compound	Phases	Temperature ⁰C	ΔH (kJ/mol)	ΔS (J/ mol/K)	ΔS/R
RN-6-2/Me	Cr- N	95.2	21.7	58.9	7.09
	N- I	172.8	0.30	0.68	0.082
	I-N	172.3	0.33	0.74	0.089
	N-SmA*	90.1	0.40	1.12	0.134
	SmA*-Bx	88.6	0.99	2.72	0.33
	Bx-Cr	65.6	13.3	39.5	4.75

6.1.3. Polarising Microscopy studies:

Polarizing optical microscopic investigations were carried out on different types of substrates viz., untreated glass plate and coverslip, nylon-66 treated glass plate, glass plates with unidirectional rubbing, commercial ITO coated cell of 5 ± 0.2μm thickness for homogeneous alignment (Instec LC3-5.0) and homeotropic alignment. The examination of the pristine samples sandwiched between a slide and a coverslip gave usual and unusual textures when examined on polarized light microscopy attached with a hot stage which are shown in **Figure 6.2** for samples **RN-6-2/Me.** The samples, on cooling from the isotropic phase, exhibited distinctly well known characteristic textures of nematic phase viz. 2- and 4-brush schlieren texture, planar and marble textures with threads as boundaries throughout the nematic phase temperature range. The growth of schlieren texture is accompanied by Brownian motion. The samples exhibited marble textures with threads in nylon-6,6 treated glass plates. **RN-6-2/Me** on further cooling from nematic phase, it undergo a phase transition from the stable nematic phase (N) spreading with ripples in the entire field of view over a period of time as shown in **Figure 6.2d.** The stable texture after cooling to the SmA* phase is shown **Figure 6.2e.** Further cooling gave a transformation in texture during transition as shown in **Figure 6.2f** resembling a characteristic texture of B2 phase. The high temperature N phase shows chiral domains [13-17] and presented in **Figure 6.3a-c.** The transition from N-SmA* is accompanied by transient Schlieren texture with undulation lines (**Figure 6.3d**). These undulation lines/fingerprint texture is stable and do not disappear over a period of time.

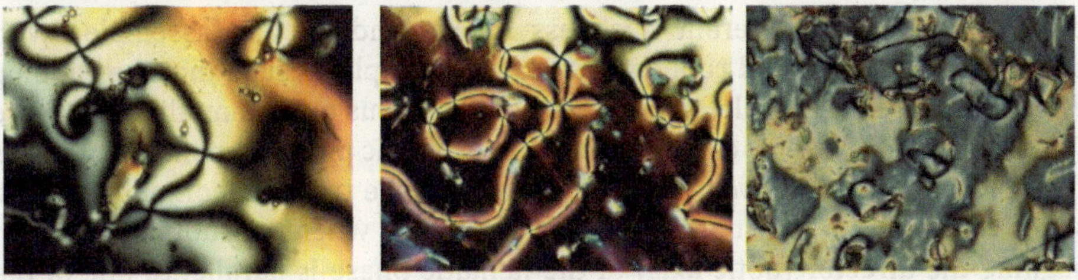

a b c

Figure 6.2(a-c): Textures of **RN-6-2/Me** containedbetween a normal glass plate and coverslip. Nematic phase at a) 172.5⁰C, b) 172⁰C and c) 91.5⁰C

d e f

Figure 6.2(d-f): Textures of **RN-6-2/Me** containedbetween a normal glass plate and coverslip. Texture during transition at d) 90.1^{0}C, e) SmA at 88.5^{0}C and f) B2 phase at 83.1^{0}C.

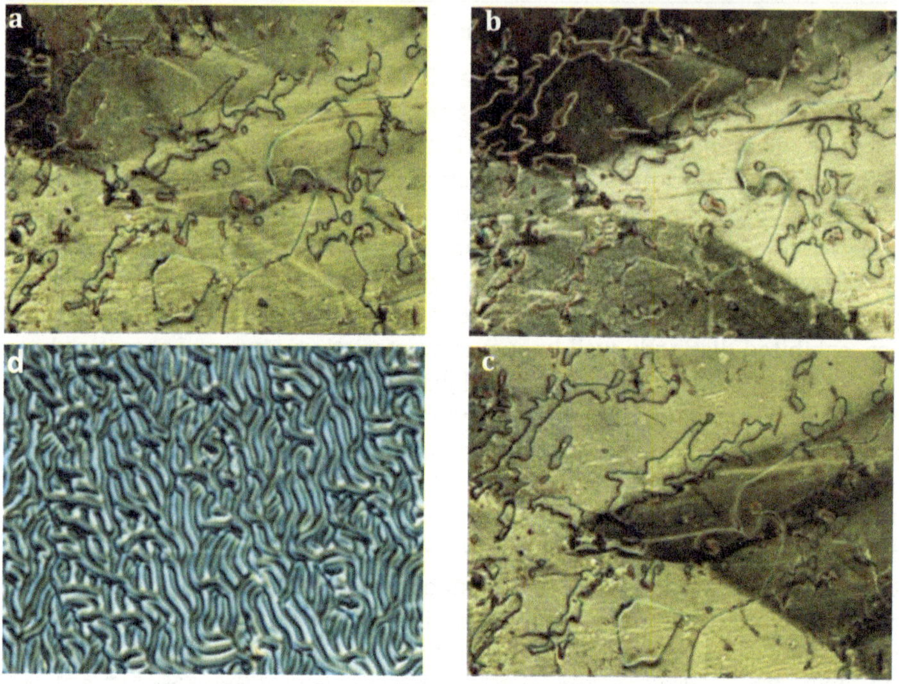

Figure 6.3: POM images of the nematic phase of **RN-6-2/Me** at 172^{0}C a) crossed polarizers, b) anticlockwise rotation of analyzer 30^0 c) clockwise rotation of analyzer 30^0 and d) Finger print texture of nematic phase in some regions of the samples preparation of **RN-6-2/Me**

Commercial cells of different thickness 5.0 μm and 9.0 μm (assembled with polyimide coated parallel rubbed ITO coated glass plates such that the rubbing directions were antiparallel to promote homogeneous molecular alignment) are filled with **RN-6-2/Me** by capillary action in isotropic phase. In nematic phase as the cell was rotated between crossed polarizers, the transmission of one of the states was found to vary uniformly from extinction, when the rubbing direction aligning the director (optic axis) of the nematic phase is parallel to one of the polarizers, to a maximum when the rubbing direction is at 45° to the polarizers. Hence the high temperature phase resembled a birefringent plate of a crystal with the optic axis in the plane of the substrate **(Figure 6.4a)**. With the variation of the sample in azimuthal angle the transmitted light intensity between crossed polarizers continuously increased to a maximum when the optic axis was oriented at 45^{0}. The nematic phase was confirmed by characteristic flickering upon tapping the sample caused by Brownian motion. On further cooling the nematic phase transformed into chiral Smectic A phase (SmA*) by a very slow growth of two types of spiral textures **Figure 6.4b** resembling the spiral filament textures of B7 phase, which later transformed into cluster like regions with complimentary colours in adjacent domains indicating opposite chirality and with a strong increase in birefringence as shown in **Figure 6.4c**. The lower temperature smectic phase exhibited two domains with uniform black back ground. Further when the polarizer and analyser are deviated clockwise or anticlockwise from crossed conditions the adjacent domains exhibited complimentary colours. These characteristic textures augment the observations of textures with untreated and nylon-66 treated cell and strongly suggest the formation of chiral domains by the generation of helical macrostructure from the twist conformation of two segments at the ester linkage. Density functional studies also confirmed such a deviation in the planes of two segments form coplanar conformation. Further cooling of the phase transformed to a stable banana phase **Figure 6.4f** via spiral domains as shown in **Figure 6.4d,e.Figure 6.4g** and f are texture observed at 50X magnification at 89 and 84^{0}C resp.

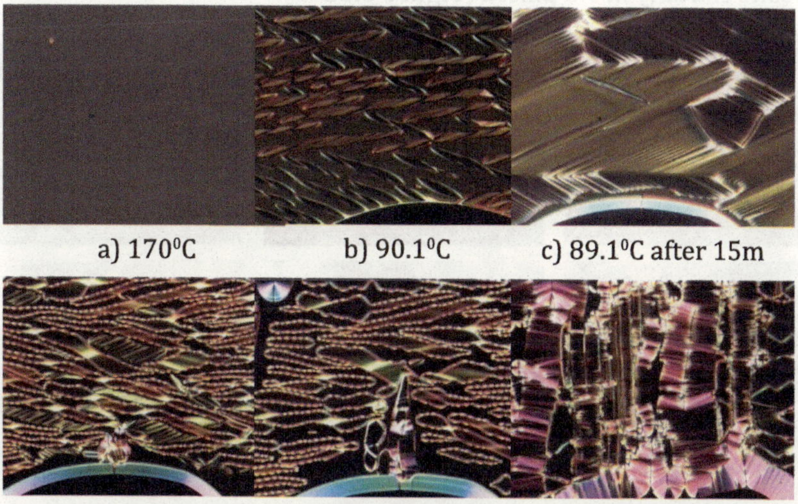

a) 170^{0}C b) 90.1^{0}C c) 89.1^{0}C after 15m

d) 89.0^{0}C e) 89.0^{0}C after 10 mins f) 85.3^{0}C

Figure 6.4(a-f): POM texture of **RN-6-2/Me** in a 5 μm planer cell. Temperature is shown below the figure.

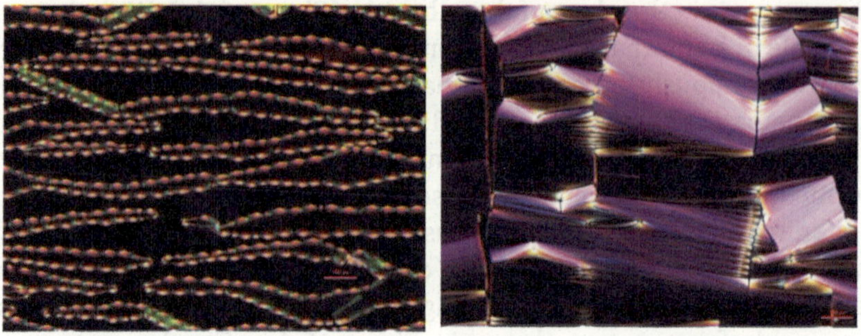

g) 89⁰C, (50X magnification)　　h) 84⁰C, (50X magnification)

Figure 6.4 (g,h): POM texture of**RN-6-2/Me** in a 5 µm planer cell under 50X magnification.. Temperature is shown below the figure.

The nematic – SmA* phase transformation is accompanied by helical filaments (**Figure 6.4b**). The spiral filaments when observed under crossed polarisers, it became apparent that they are found to be of two different types with exactly opposite colour dispersion on rotation of the analyser. The observation of areas of opposite handedness as evident by the pink and green complimentary coloured domains (**Figure 6.5A-C)** is a signature of helically twisted superstructure with opposite twist in the two domains with opposite handedness. Further the birefringent planar domain texture of the fluid nematic phase (N) exhibit regions of opposite chirality in 9.0 µmcell (**Figure 6.5D-F)**.

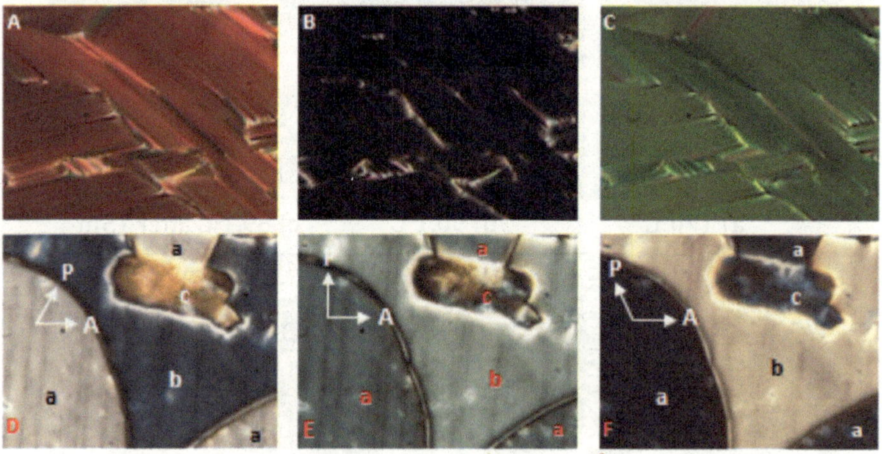

Figure 6.5: Texture of the compound **RN-6-2/Me**at 84.5⁰C under cross polarizer (B) and uncrossing polariser 10⁰ clockwise (A) and anti-clockwise (C). Second row are textures of homogeneously aligned 9 µm cell at 89⁰C under crossed polarizers (E), and slightly uncrossing the analyzer with respect to polarizer towards clockwise (D) and anticlockwise (F). White arrows show the direction of polarizer and analyzer

This optical property has been observed in several bent core materials being characteristic of spontaneous resolution of helical chiral domains of opposite handedness. The DSC spectra could not detect phase transitions.

It is apparent from the appearance of the chiral filaments in 5 μm and 9 μm cells that the pitch of the twisted superstructure in the domains of opposite handedness is clearly smaller than the sample thickness and the texture and colours observed under crossed polarizers are the textures and birefringence colours of a nematic phase. Moreover, the two chiral filament structures observed at the N-SmA* and SmA*-B_X phase transformations and in the SmA* phase in these four ring-achiral bent-core unsymmetrical materials on uncrossing the polarizers are clearly different in appearance from surface induced twist domains which can occur in any nematic material. Firstly, the observed domains in untreated thin films of pristine samples are not fixed which suggest that the place they occur is not controlled by the surface. Secondly, the growth of two stable domains in the thin film cells with equal probability suggests the degeneracy of opposite helical twist directions. Thirdly, the appearance of opposite colour dispersion of complimentary colours on equal angles of rotation of the analyser from the crossed position suggests that the pitch of the helically twisted superstructure in the two areas is not only of opposite sense but also of the same magnitude.

The polarising optical texture of compound **RN-6-2′Me** is also observed on a homeotropically aligned cell. The nematic phase is observed as uniform slightly blackish (ash colour). On cooling, smectic phase grows dendritically and covers the viewing area completely. This SmA* phase remains for a long temperature range. On further cooling *striated texture* can often be observed. This is sometimes referred to as myelinic or **chevron** texture, [18,19] with a line pattern of transition bars. The general structure of horizontal chevrons is in principle similar to the striped bookshelf texture, which in observed for SmC* materials. **Figure 6.6b** shows typical zigzag defects of a surface stabilized SmA* phase in the (vertical) *chevron* geometry. At the zigzag lines, domains of opposite layer inclination meet. The propagation of horizontal chevron lines are shown in **Figure 6.6c.**Texture of tilted smectic phase with completely filled *horizontal chevron* structure is shown in figure d. The helix lines clearly indicate the direction of the smectic layer planes, and in the region where two domains of opposite layer tilt meet, a vertical chevron defect forms the domain boundary. Chiral nature of SmA phase is observed by slightly uncrossing the polarizer.

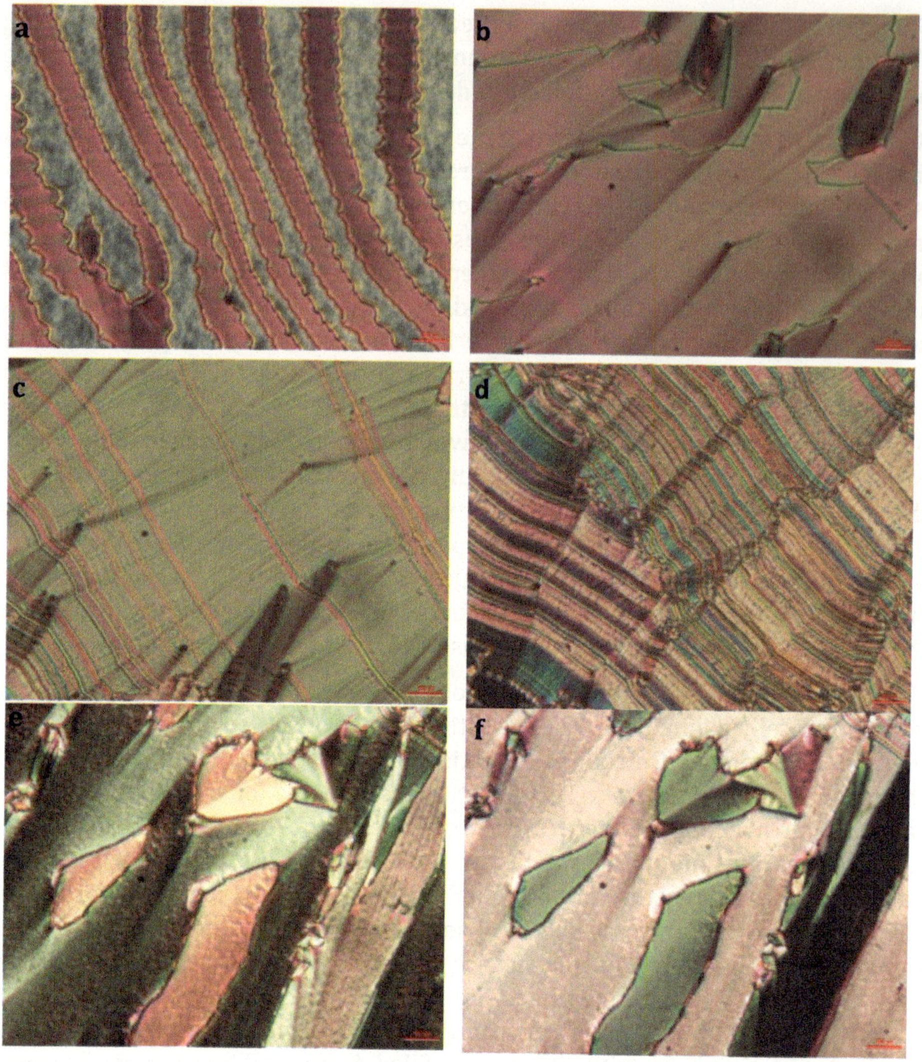

Figure 6.6: Texture of RN-6-2/Me on a homeotropic cell. a) growth of SmA from N phase at 90°C, b) SmA phase with vertical chevron lines, at 87°C c) propagation of horizontal chevron line at 83.3°C and d) tilted smectic phase with chevron lines at 77°C. e) and f) are chiral domains of SmA phase observed by rotating polariser clockwise and anti-clockwise respectively at 83°C.

The transition temperatures, enthalpies and entropies associated with the phase transitions as obtained from differential scanning calorimetry for **RN-6-2/Me** at different scan rates in the cooling cycle are presented in **Figure 6.7**. The DSC performed at different scanning rates clearly documented a change in the appearance of the N–SmA* and SmA*- Bx phase transitions. After two cycles of heating and cooling the two transition peaks smeared into one peak, thereby relegated the N- SmA* transition in obscurity and to assign it as single thermal event. On further investigation of the mesophase behaviour of the material by the POM studies we observed unusual behaviour in the texture patterns dependent on the thickness of film. In a 5 μm cell the N-SmA* transition is accompanied by the spiral filament textures during the transformation which takes place with longer time durations for the growth of spiral texture, which subsequently disappeared after third heating and cooling cycles because of prolonged heating and cooling cycles. However the spiral growth texture at the SmA*-B2 phase transformation appeared even after repeated cycles of heating and cooling.

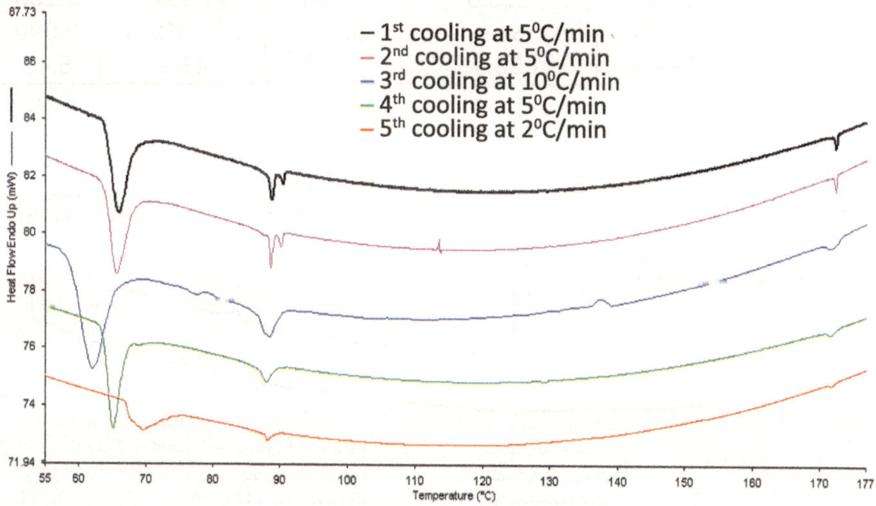

Figure 6.7: Thermal history of **RN-6-2′Me** at different cooling rates

6.1.4 Mesomorphic property of RN-8-2′Me,

a) DSC study:

The compound exhibited three enantiotropic phases in the heating and cooling cycles with low transition enthalpies above the melting of the sample. The phase transition temperatures, enthalpies and entropies associated with the phase transitions as obtained from DSC for **RN-8-2′Me** at 5^0 min^{-1} are presented in **Table 6.3** and **Figure 6.8.**

Table 6.3: Transition temperatures obtained from differential scanning calorimetry at a scan rate of 5°C/min in 2nd heating and cooling data for **RN-8-2'/Me**

Compound	Phases	Temperature °C	ΔH (kJ/mol)	ΔS (J/mol/K)	ΔS/R
RN-8-2'/Me	Cr- Bx	96.7	23.12	62.7	7.54
	Bx-N*	100.3	0.31	0.83	0.099
	N*-N	138.8	0.74	1.796	0.216
	N- I	162.4	0.325	0.748	0.089
	I-N	161.3	0.346	0.797	0.096
	N-N*	135.3	0.554	1.357	0.163
	N*-Bx	96.2	0.276	0.748	0.090
	Bx-Cr	58.9	14.99	45.1	5.43

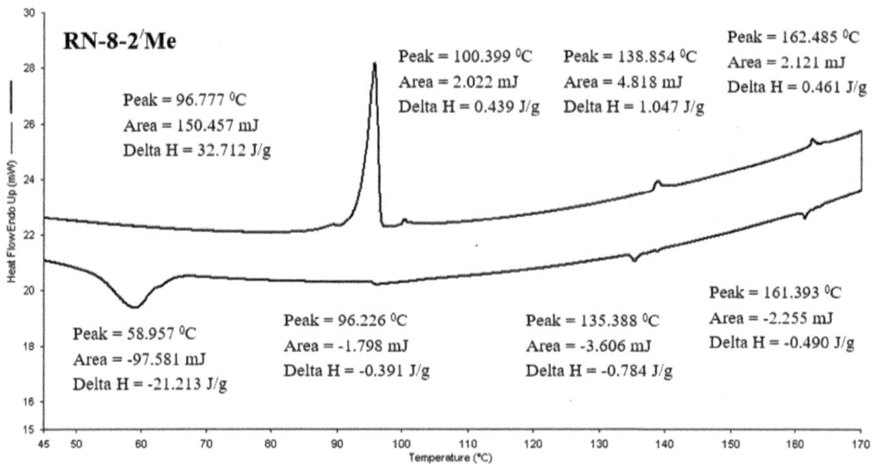

Figure 6.8: Differential Scanning Calorimetric thermogram of **RN-8-2'Me** for second heating and cooling at scan rate 5 °C /min.

DSC spectrum of **RN-8-2'Me** shows four transitions at 96.7°C (ΔH = 23.1 kJ/mol, ΔS = 62.7 J/mol/K), 100.3°C (ΔH = 0.31. kJ/mol, ΔS = 0.83 J/mol/K), 138.8°C (ΔH = 0.74 kJ/mol, ΔS = 1.79 J/mol/K) and 162.4°C (ΔH = 0.32 kJ/mol, ΔS = 0.75 J/mol/K) in heating cycle and four transitions at 161.3°C (ΔH = 0.35 kJ/mol, ΔS = 0.79 J/mol/K), 135.3°C (ΔH = 0.55 kJ/mol, ΔS = 1.35 J/mol/K), 96.2°C (ΔH = 0.27 kJ/mol, ΔS = 0.75 J/mol/K) and 58.9 °C (ΔH = 15 kJ/mol, ΔS = 45.1 J/mol/K) in the cooling cycle.

b) POM studies:

The compound **RN-8-2'Me** is observed under microscope placed between normal glass plate-coverslip and planer cell (3.5μm, **Figure 6.9a-f**). The normal plate shows schlieren textures with two and four brush disclination (a). On cooling, the birefringence decreases and become optically isotropic (b). In such phases the molecules become perpendicular to the glass substrate. On further cooling, focal conic textures (c) of smectic phase grows as betonnets from black region. This phase is stable for long temperature range. On cooling this transform to another smectic phase (d) before crystallisation (e). **Figure 6.9f** shows undulation lines of Smectic phase

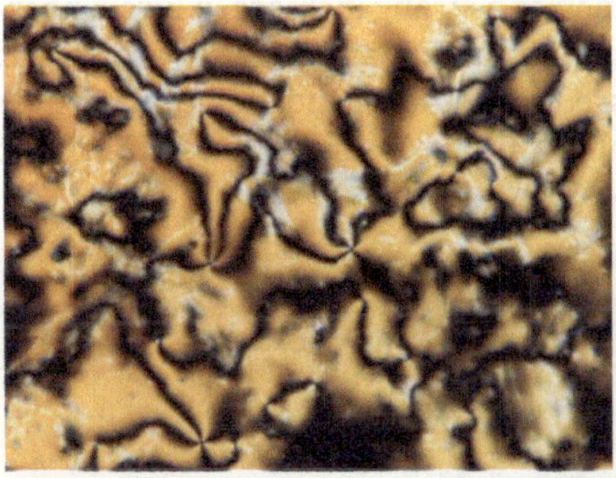

(a) 159.3°C, 2-and 4-brush defects

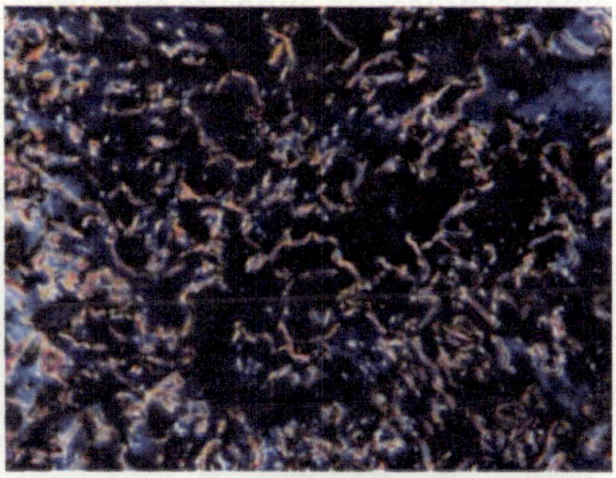

(b) 128.8°C

(c) 124.8°C

(d) 82°C

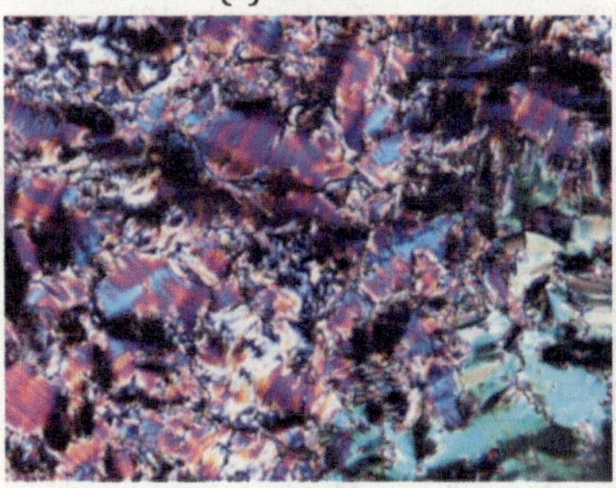

(e) 79°C

(f) 132°C

Figure 6.9: POM textures of **RN-8-2′Me** on a glass plate, cover slip. Temperature are presented below the image.

Planar oriented cell of **RN-8-2′Me** is observed under polarizing microscope. Nematic phase is observed as uniform black colour when rubbing direction is parallel to one of the polariser direction. On changing the rubbing direction by rotating sample stage, the birefringence and colour changes equally on either side of rotation. On further cooling the nematic phase transformed into chiral Smectic A phase (SmA*) by a very slow growth of two types of spiral textures **Figure 6.9b** resembling the spiral filament textures of B7 phase, which later transformed into cluster like regions with complimentary colours in adjacent domains indicating opposite chirality and with a strong increase in birefringence as shown in **Figure 6.9d-f**. Further when the polarizer and analyser are deviated clockwise or anticlockwise from crossed conditions the adjacent domains exhibited complimentary colours. On cooling the phase parallel lines of horizontal chevron geometry covers the fan (**Figure 6.9c**).

a) 147^{0}C, nematic phase (N)

b) N-SmA* phase transition at 135.1^{0}C

c) 91^{0}C horizontal chevron lines

d) domains of SmA at 135^{0}C under cross polariser

e) clockwise rotation of polariser by 10^0

f) anti-clockwise rotation of polariser by 10^0

Figure 6.9: Textures of **RN-8-2/Me** on a 3.5μm planar oriented cell.

Transition Temperatures, Enthalpies and entropies other compounds of other homologue of the homologous series **RN-n-2Me,** (n=10, 12, 14, 16, 18) are presented in **Table 6.4**.

Table 6.4: Transition Temperatures (^{0}C), Enthalpies (*Italic*,kJ/mol) and entropies (*Italic*, J/mol/K) of homologous series **RN-n-2Me**.

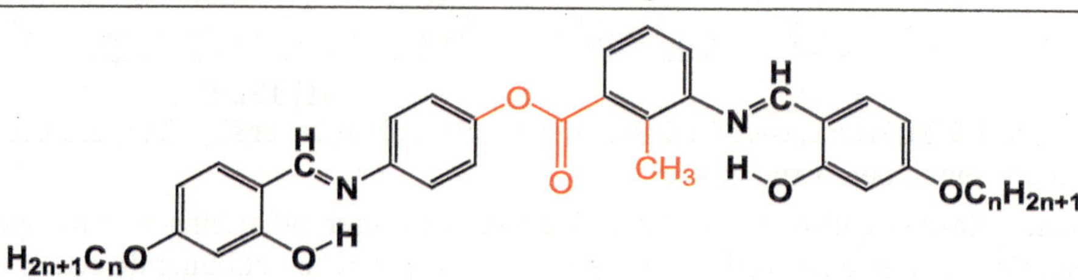

Compound	R	Cr	Heating Cooling	SmC	Heating Cooling	I
RN-10-2/Me	$C_{10}H_{21}$	•	96.50 *[28.9, 78.3]*	•	160.94 *[5.0, 11.4]*	•
		•	61.98 *[18.8, 56.3]*	•	159.63 *[4.2, 9.8]*	•
RN-12-2/Me	$C_{12}H_{25}$	•	96.82 *[31.5, 85.2]*	•	155.37 *[4.3, 10.1]*	•
		•	65.41 *[6.7, 19.8]*	•	153.46 *[5.2, 12.1]*	•
RN-14-2/Me	$C_{14}H_{29}$	•	98.49 *[54.4, 146.6]*	•	163.58 *[9.3, 21.2]*	•
		•	71.43 *[36.3, 105.5]*	•	162.49 *[10.1, 23.1]*	•
RN-16-2/Me	$C_{16}H_{33}$	•	101.08 *[52.3, 139.9]*	•	161.95 *[7.7, 17.3]*	•
		•	75.61 *[38.7, 110.9]*	•	160.42 *[7.0, 16.2]*	•
RN-18-2/Me*	$C_{18}H_{37}$	•	130.05	•	134.34	•
		•	111.22	•	133.15	•

* Transition temperatures of RN-18-2/Me are from microscopic observation

6.1.5. Compound RN-10-2/Me:

a) DSC studies:

The differential scanning calorimetric thermogram of the compound **RN-10-2/Me** is presented in **Figure 6.10. RN-10-2/Me** exhibited two transitions in the heating cycle and melts at 96.5°C (ΔH = 28.9 kJ/mol, ΔS = 78.3 J/K/mol) and transforms into the isotropic phase at 160.9°C (ΔH = 5.0 kJ/mol, ΔS = 11.4 J/K/mol). In the cooling cycle also it exhibited enantiotropic phase transitions at 159.6°C (ΔH = 4.2 kJ/mol, ΔS = 9.8 J/K/mol) and 61.9°C (ΔH = 18.8 kJ/mol, ΔS = 56.3 J/K/mol). The compound exhibits wide mesomorphic range of ~98°C in the cooling cycle.

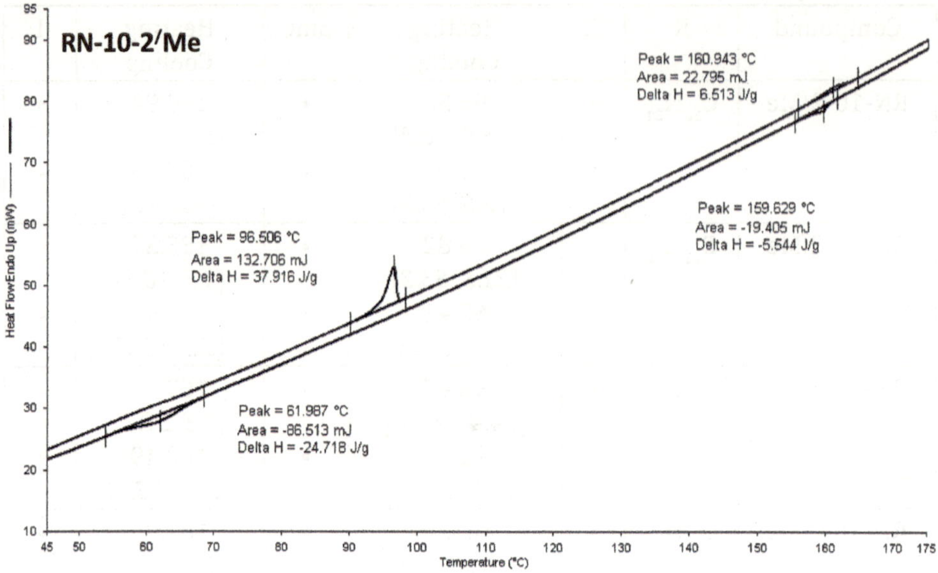

Figure 6.10: DSC spectrum of **RN-10-2/Me** at a heating rate of 5°C/min in second heating and cooling

b) POM studies:

The optical textures observed in smectic C and B2 phases and during the phase transitions are displayed in **Figure 6.11.** On cooling the isotropic liquid of the sample growth of schlieren texture with undulated lines appeared indicating the phase as SmC. In a commercial polyimide coated anti-parallel rubbed cell filled with the sample, when it is cooled from the isotropic phase broken focal conic fan domains appeared indicating the phase as SmC. Further cooling of the sample revealed no change in texture in appearance but transient transition bars are observed at 96.0 °C which are observed for B2 phase. DSC could not detect this phase transition.

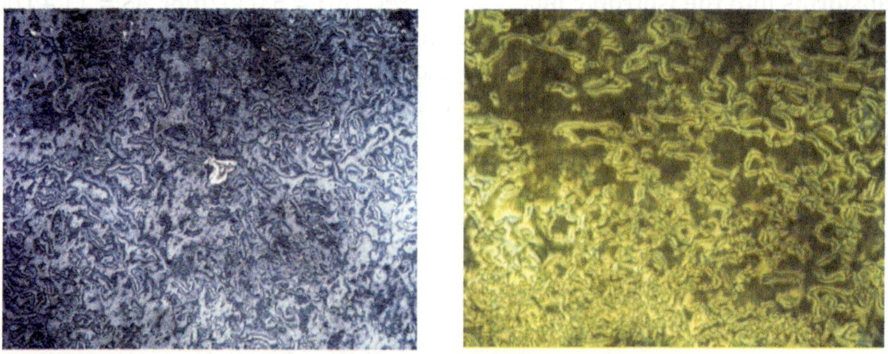

a) 158.1°C b) 149.7°C

c) 151.0°C

d) 96.0°C

e) 95.3°C

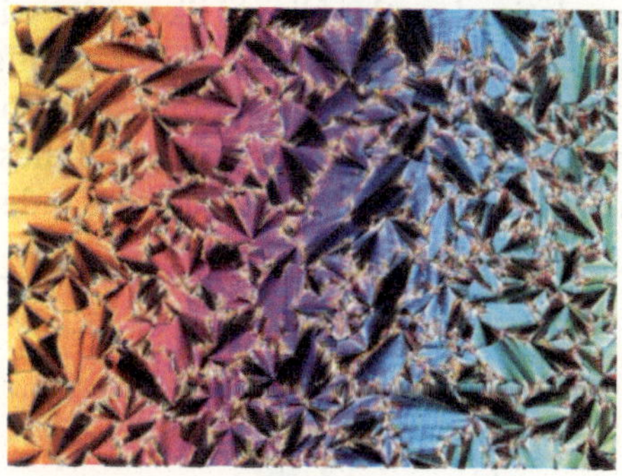

f) 95.3°C at different site.

Figure 6.11: Polarizing optical microphotographs observed in **RN-10-2′Me** in a thin film in a)and b) schlieren textures (normal glass plate and coverslip). Textures observed in polyimide coated cell of 5.0 μm thickness with antiparallel alignment. c) broken focal conic fan textures of SmC phase d) transient transition bars across fans e) paramorphotic texture of B2 Phase and f) B2 phase at different site.

c) X-ray diffraction studies:

The XRD studies carried out on the mesophases of the compound **RN-10-2′Me** are described below. The compound **RN-10-2′Me** exhibits one sharp reflection in the small angle region **(Figure 6.12)** at d_1 = 41.65 Å, corresponding to the layer thickness at a temperature of 150°C. This reflection indicates a lamellar ordering in the mesophase. Further, the first order reflection at 41.65 Å can be indexed to a reflection from (0 1) plane which corresponds to nearly the molecular length of 48.97 Å by assuming a fully extended all trans-conformation of the alkoxy chain **(Figure 6.13)**. This clearly shows that the bent-core molecules are tilted in the mesophase corresponding to a tilt angle of 31.7°. Also, a diffuse peak in the wide angle region at 4.72 Å indicates the fluidity in the layers and absence of an in-plane order. The mesophase exhibits a broken focal-conic texture and also a schlieren texture, inferring the possibility of smectic C phase. On further cooling the sample exhibited a phase transition at 96.1°C with the appearance of transient transition bars on the focal conic fans which disappeared at 96.0°C after allowing the sample for a sufficient time 3 ~ 4 hours. Further it cannot be aligned homeotropically and hence the possibility of any orthogonal phases has been eliminated. XRD pattern in the mesophase at 85°C exhibited a strong reflection in the small angle region (42.33Å) and its second order reflection (21.17 Å) are characteristics for a smectic layer structure. Further the broad diffuse scattering has a maximum at 4.23Å indicates a liquid-like order within the smectic layers. The temperature variation of layer thickness has been illustrated in **Figure 6.14**. In principle the X-ray investigations are also compatible with a smectic C phase but the results of texture transformation, electro-optical measurements, the width of transversal correlation length of liquid like order within the layers unambiguously point out that the low temperature phase is different from high temperature smectic C phase. In addition, the mesophase can be switched electro-optically and exhibits an antiferroelectric phase. From all these observations, it can be concluded that this low temperature mesophase is indeed a $SmCP_A$ (B_2 family).

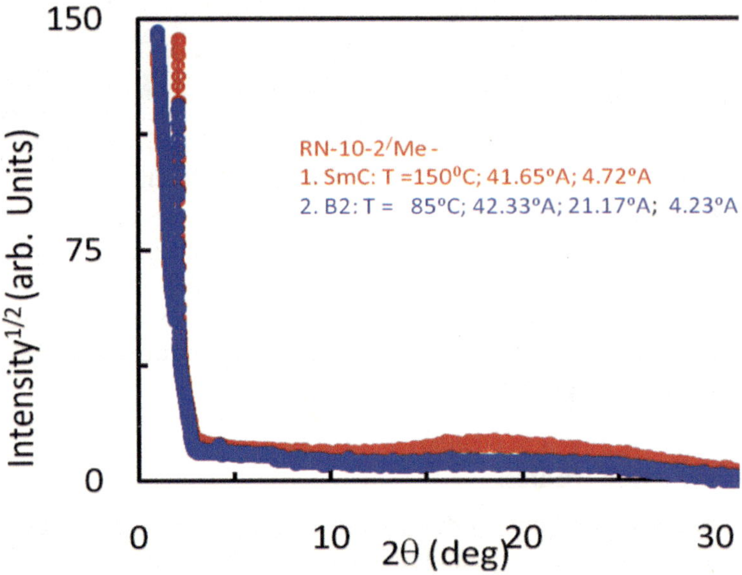

Figure 6.12: X-ray diffraction pattern of **RN-10-2/M** at 150 and 85°C.

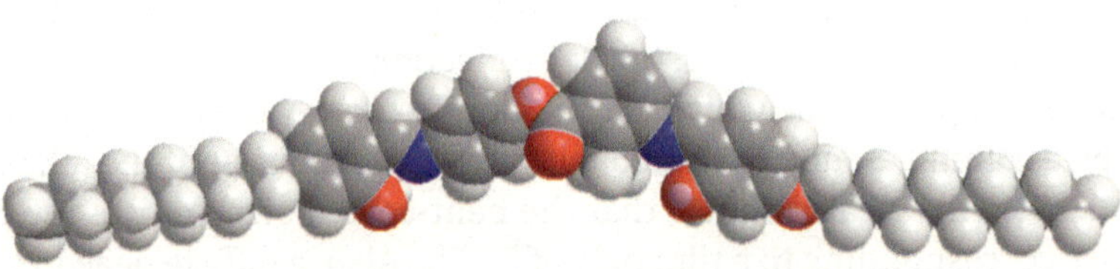

Figure 6.13: Fully extended all trans conformation of **RN-10-2/Me**. (L = 48.97Å)

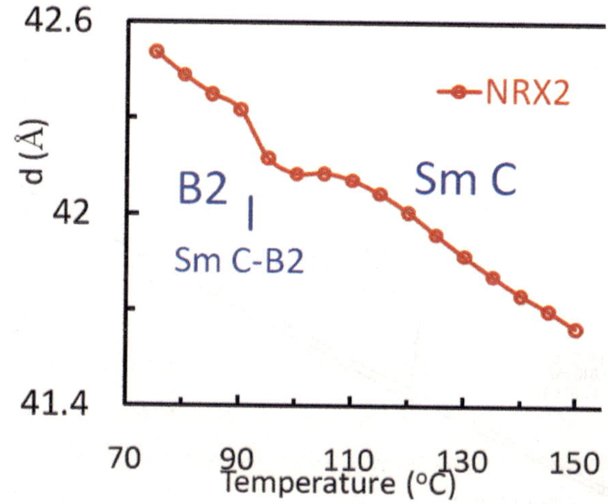

Figure 6.14: Temperature variation of layer thickness with temperature in SmC and B2 phases.

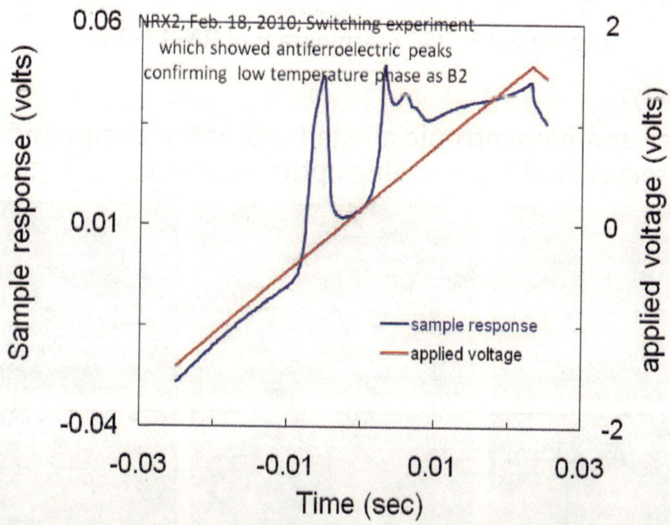

Figure 6.15: Current response to triangular voltage indicating antiferroelectric switching behaviour in the SmCPA phase (B2 family) of compound **RN10-2/Me** (T = 85°C)

6.1.6 Compound RN-14-2/Me:

a) DSC study:

The compound **RN-14-2/Me** exhibit one enantiotropic phase transition. Transition temperature, enthalpy and entropy associated with phase transition are presented in **Table 6.4.** the DSC thermogram is presented in **Figure 6.16**.

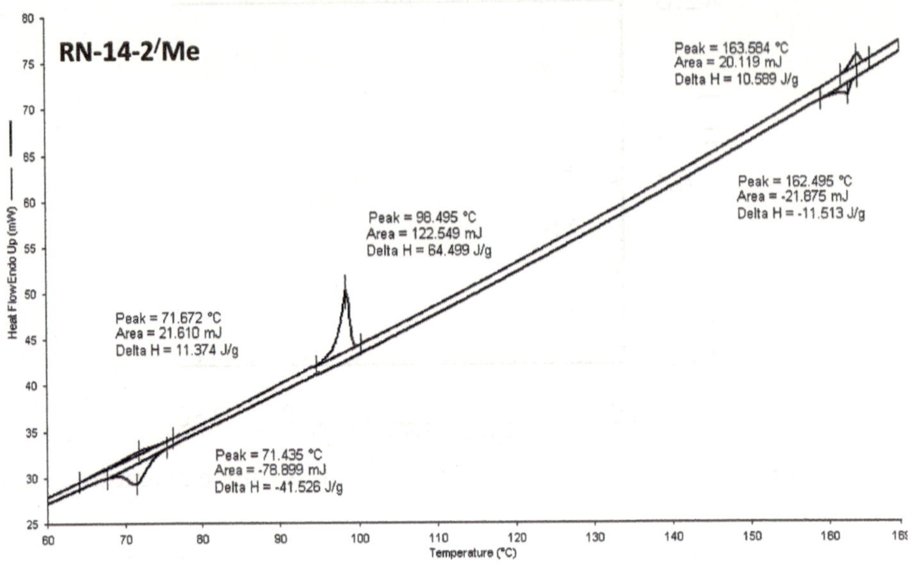

Figure 6.16: DSC thermogram of **RN-14-2/Me.**

b) POM study:

Both planer and homeotropic aligned cell of the compound **RN-14-2/Me** are observed under polarising optical microscope (**Figure 6.17**).

Figure 6.17: POM textures of **RN-14-2/Me** on a planar oriented surface (Nylon 66) at 140, 100.3, 99.7 °C (1st row, left to right) and on a homeotropic (HTAB) cell at 139, 99 and 93.8°C (2nd row, left to right).

In both the cell similar textures were observed with slight variation in transition temperature. On cooling the isotropic melt, schlieren texture with four brush defects are observed. On cooling the birefringence decreases and the haziness increases. This types of textures are observed for anticlinicSmC phase. Before the crystallisation the birefringence suddenly increases. [20]

c) X-ray diffraction study:

The powder X-ray diffraction patterns of the liquid crystalline phase of the compound **RN-14-2/Me** at different temperature are displayed in **Figure 6.18a-e**

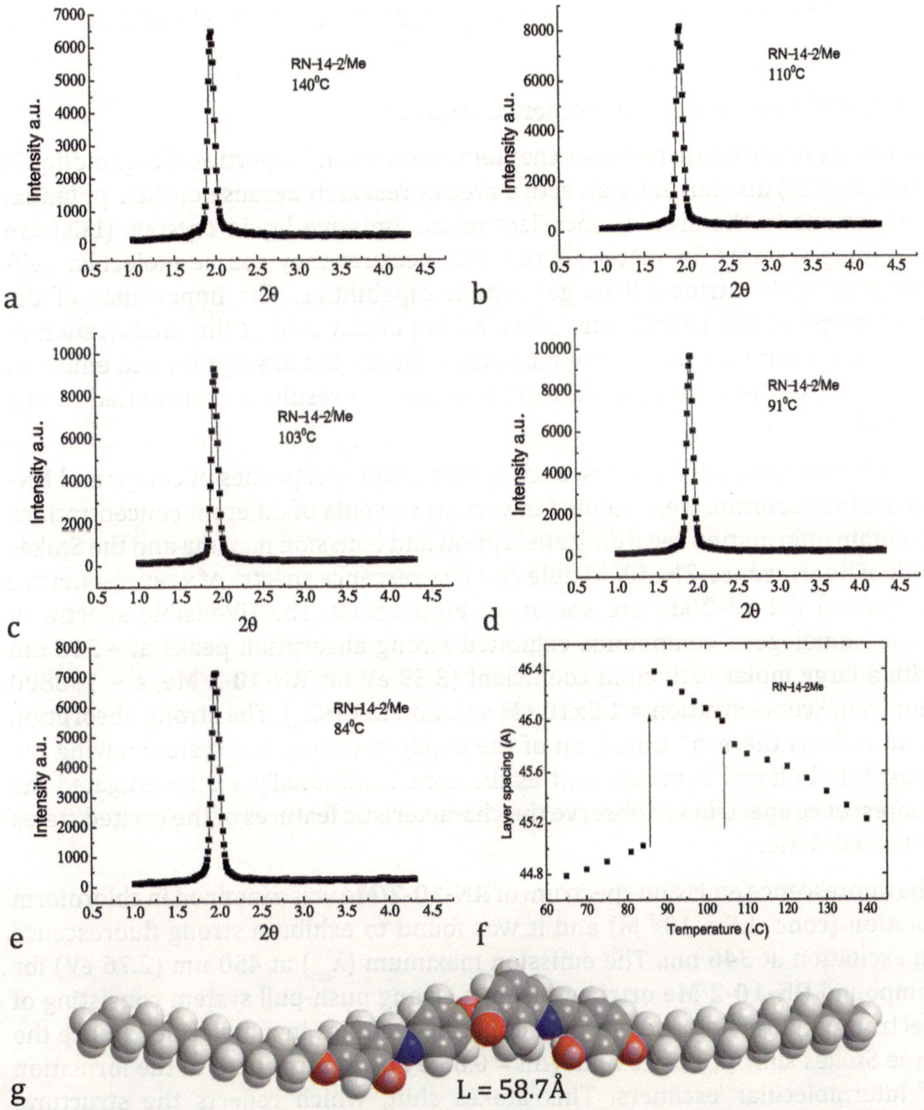

Figure 6.18: X-ray diffraction pattern at different temperature (a-e), Variation of layer thickness with temperature (f) and fully extended all trans conformation (g) of **RN-14-2/Me.**

In the small angle region compound **RN-10** exhibited one sharp layer reflection with a spacing of 45.2 Å (T = 140°C) which is smaller than the length of the molecule in the most extended conformation (L = 58.7Å, **Figure 6.18g**) inferring that the molecules are arranged in tilted fashion with a large tilt angle ($\theta = 39^0$)

in smectic layers (SmC). The variation of layer spacing with temperature is presented in **Figure 6.18f.** The graph shows a gradual increase in layer thickness with decrease in temperature till the temperature T = 86°C with layer thickness 46.4Å with a jump at 105°C. At 86°C the layer thickness suddenly decreases to 45Å. This two temperature corresponds to SmC-SmC_{anti} (105°C) and SmC_{anti} - Cr (86°C).

6.1.7. UV-Visible and fluorescence studies:

As already described in previous chapter the emission properties of organic liquid crystals (LCs) are currently an active area of research because of their potential applications in the area of optoelectronics. Emissive liquid crystals (LCs) are intriguing systems for materials research because they couple molecular self-assembly with intrinsic light generation capabilities. The importance of the substituent on the central core plays an important role in the modification of electronic characteristics of the molecules. Hence the absorption and emission properties of the compounds are studied and the results are presented in this section.

The UV absorption and fluorescence spectroscopic properties of compound **RN-10-2/Me** in solution were studied in various solvents of different concentrations to obtain information regarding absorption and emission maxima and the Stokes shift of fluorescence. The UV-Visible and Fluorescence spectra of a representative compound **RN-10-2/Me** are shown in **Figure6.19**. The UV-visible spectra in these homologous compounds exhibited strong absorption peaks at ~346 nm with a large molar extinction coefficient (3.59 eV for **RN-10-2/Me,** $\varepsilon \sim 198800$ $Lmol^{-1}cm^{-1}$,concentration = 1.5×10^{-5} M solution in $CHCl_3$). The strong absorption band reflects the π-π* transition of the highly π-conjugated system having the substituted phenyl benzoate unit as the core. Additionally, we investigated the fluorescence spectrum to observe the characteristic features of the excited states of **RN-10-2/Me.**

The fluorescence emission spectrum of **RN-10-2/Me** was examined in chloroform solution (conc. 1.5×10^{-5} M) and it was found to exhibit a strong fluorescence on excitation at 346 nm. The emission maximum (λ_{em}) at 450 nm (2.76 eV) for compound **RN-10-2/Me** originate from a strong push-pull system consisting of electron withdrawing and electron donating moieties in a molecule. Hence the large Stokes shift (λ_s) of 104 nm (ΔE = 0.83 eV) was attributed to the formation of intermolecular excimers. This Stokes shift, which reflects the structural relaxation of the excited molecule, is significantly larger than in reported push-pull systems exhibiting liquid crystal behaviour [21-24] indicating that the molecular conformation changes upon excitation. These results are in good agreement with the reported results of molecular J-aggregates [25] in which the excitonic energy is delocalized as a result of intermolecular coupling within the head-to-tail arrangement of the molecules in the solution. The optical band gap investigations of thin films in LC phase and solid state for their possible applications and Fluorescence confocal polarizing microscope studies to detect the polarization dependent fluorescence in mesophase of these compounds are currently in progress.

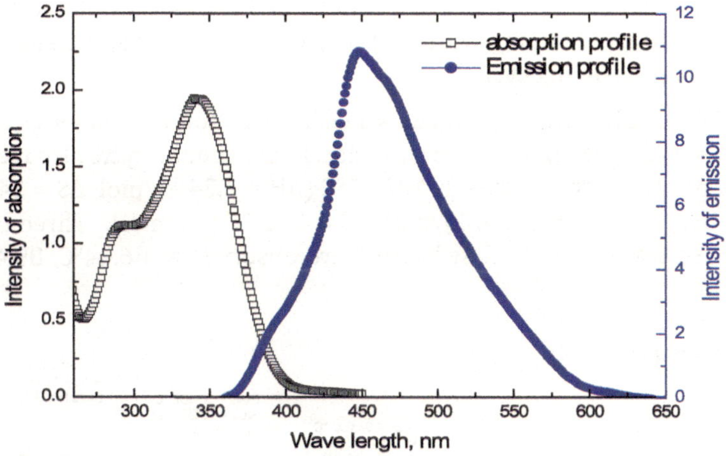

Figure 6.19: UV-Visible and fluorescence spectra of **RN-10-2/Me** in 1.5×10^{-5} M chloroform solution.

6.2. Compound RN-14-4/Me:

[4-(N-4-alkyloxysalicylidenemino)phenyl]-[5-(N-4-alkyloxysalicylidenemino)-4-methylbenzoate]:

6.2.1. Synthesis:

The synthesis of **RN-14-4/Me** is carried out using the same procedure as described in 6.1.1. 4-methyl-5-nitrobenzoic acid is converted to its acyl chloride by the reaction of thionyl chloride. This acyl chloride is condenced with 4-nitrophenol to get dinitro product which on reduction gives the diamine central core. The terminal decyloxy salicylaldehyde is synthesised by Williamson etherification of 2,4-dihydroxybenzaldehyde with n-decyl bromide. Schiff's base condensation of the diamine core and decyloxy salicylaldehyde is carried out in absolute ethanol in presence of few drops of acetic acid as catalyst. The product is filtered off in warm condition and recrystallised several times. The detailed synthetic procedure is given in Chapter 3. The formation of all of the compounds was confirmed by ^{1}HNMR and IR spectroscopy. The purity of the sample was established by elemental analysis and found to be in good agreement with the theoretical values within the experimental error. The details of the experimental procedures along with the spectroscopic data for compound are presented in the experimental part of **Chapter 3**. The liquid-crystalline behaviour of the synthesised compound had been investigated by optical microscopy, differential scanning calorimetry and x-ray diffraction technique.

6.2.2. Mesomorphic properties, DSC and POM investigations:

a) DSC studies:

DSC thermogram of compound **RN-14-4/Me** is presented in **Figure 5.20.** Graph shows enantiotropic phase transitions. **RN-14-4Me** exhibited five phase transitions in heating among which three are crystal to crystal transitions and two transitions in cooling cycle.

Crystal 98.30⁰C (ΔH = 3.70 kJ/mol, ΔS = 117.22 J/K/mol) LC 103.49⁰C (ΔH = 3.45 kJ/mol, ΔS = 27.75 J/K/mol) Isotropic phase. In the cooling cycle also it exhibited enantiotropic phase transitions at 102.73⁰C (ΔH = 3.24 kJ/mol, ΔS = 34.51 J/K/mol) and 89.21⁰C (ΔH = 27.74 kJ/mol, ΔS = 111.77 J/K/mol). Three crystal to crystal phase transitions in first heating are observed at 86.44⁰C, 89.46⁰C and 97.55⁰C.

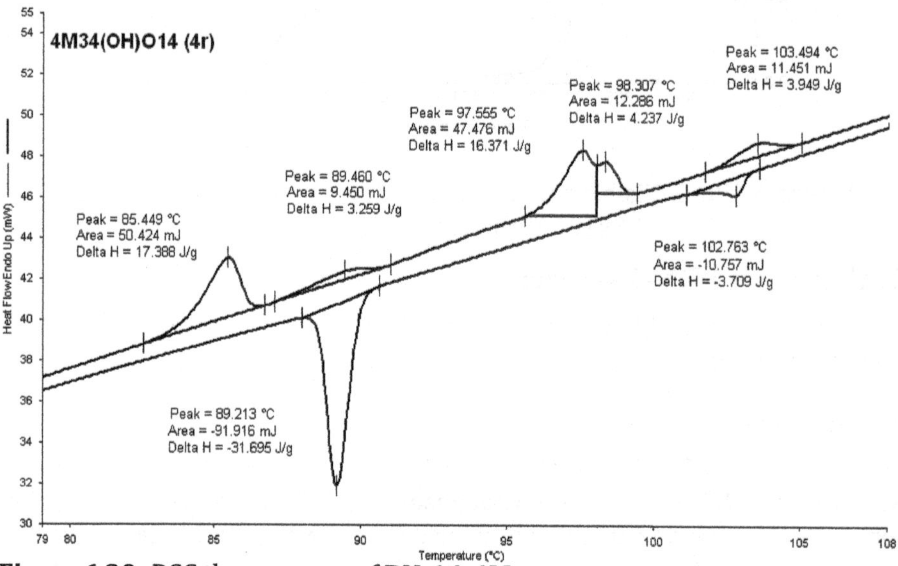

Figure 6.20: DSC thermogram of **RN-14-4Me**

b) POM studies:

On cooling the compound (**RN-14-4/Me**) SmA like batonnets and focal conic fan textures (**Figure 6.21**) appeared from the isotropic melt at 102.5⁰C. On further cooling, at 96.3⁰C the focal conic texture is masked by transition lines like indicating a transition to Bx phase. However there is no signature of this SmA-Bx phase transition in DSC spectrum. On further cooling the broken fan texture disappeared slowly become optically isotropic in the entire field of view to form a dark conglomerate phase. This optically isotropic phase shows an additional remarkable feature called macroscopic chirality, which can be seen by slightly uncrossing the polarizers. Under these conditions a domain structure composed of dark and bright regions can be observed. Reversing the direction of polarizer and analyzer reverses the brightness of these domains. This is an indication of a mesophase structure composed of a conglomerate of macroscopic domains of opposite handedness with chiral behaviour [26, 27]. It was proposed that this optical activity is due to the inherent chirality of the polar SmCP layers [28] but it can only be observed if the phase structure is homogeneously chiral, i.e., for a SmC_AP_A or SmC_SP_F structure, and if, in addition, the birefringence of the mesophase itself is sufficiently low. The low birefringence is thought to result from a sponge-like deformation of the layers [28, 29].

Figure 6.21: (a) Focal conic texture of SmA phase at 96.3 ⁰C, (b) texture of B_X phase transition (c, d) transition from B_X to an optically isotropic phase at 88.3 ⁰C, (g, h) Texture of the optically isotropic phase /SmCP* phase (dark conglomerate phase) of compound **RN-14-4Me** at 86 °C between slightly uncrossed polarizers (8⁰). The brightness of the domains is exchanged by uncrossing the polarizers in either one or the other direction, indicating regions of opposite chirality sense. White arrows indicate the direction of polariser and analyser.

6.2.3. X-ray diffraction studies:

To confirm the mesophase structures X-ray diffraction studies are performed on a representative compound belonging to the homologous series **RN-14-4/Me**. X-Ray measurements were performed on unaligned (powder) samples in a 0.5 mm diameter Lindeman capillary tube (sealed at both ends) with Cu Kα (λ = 1.5418Å) radiation using the P Analytical X'Pert PRO MP X-ray diffractometer consisting of a focusing elliptical mirror and a fast resolution detector (PIXCEL). A typical X-ray diffraction patterns obtained for the compound **RN-14-4/Me** at 100 °C and 92 °C are shown in **Figure 6.22.** The diffraction pattern indicated a strong reflection at 2θ = 2.00° corresponding to 43.96Å. This corresponds to the molecular length n which is slightly lower than the estimated molecular length of 57.4 Å in all trans conformation. The POM texture of focal conic fans confirmed the SmA phase and hence the end aliphatic chains must have interdigitated to explain the phase structure. In addition to these reflections a broad diffuse wide angle reflection at 2θ = 19.3° corresponding to 4.56 Å reflects the spacing between the aliphatic chains indicative of a liquid-like in-plane order of the fluid phase. The diffraction pattern at 92°C recorded a strong reflection at 2θ = 1.98° corresponding to 44.96 Å. Further a broad diffuse wide angle reflection at 2θ = 19.3° corresponding to 4.56 Å reflects the spacing between the aliphatic chains indicative of a liquid-like in-plane order of the fluid phase. The layer thickness and full width at half maximum as a function of temperature are plotted in **Figure 6.23.** Below T = 95 °C, the diffraction peak becomes much broader and the position of this peak does not change as the temperature decreases to room temperature. This kind of behaviour is observed for other materials having B4 phase. For other B4 phase materials, like NOBOW, usually a few peaks in the wide angle can be seen. Based on the POM texture, X-ray studies and large enthalpy values associated with the phase transition the mesophase in this compound is identified to exhibit Iso-SmA-B2-DC phase-Cr phase variant.

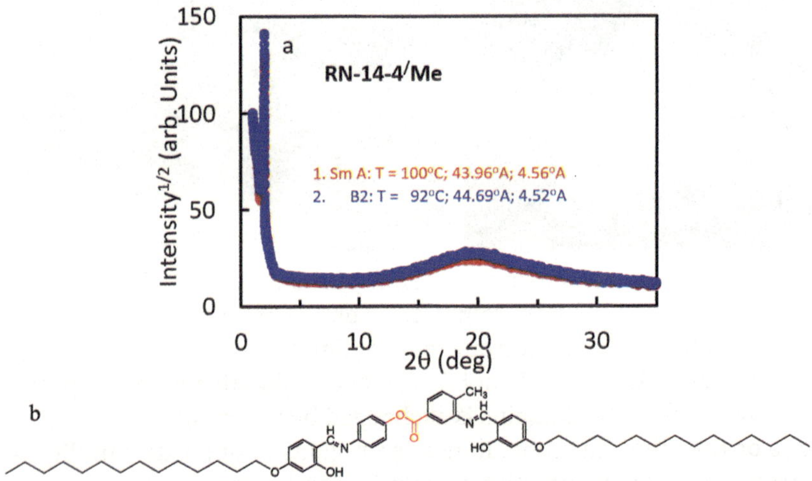

c

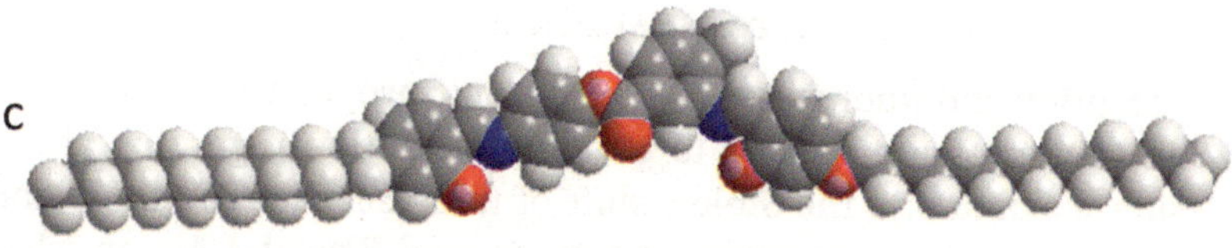

molecular length = 57.4 Å (theoretical)

Figure: 6.22:. a) X-ray diffraction pattern at T = 100°C and 92°C.b) Molecular structure and c) Space filling model of energy minimized structure of the molecule **RN-14-4/Me** with fully extended end chains.

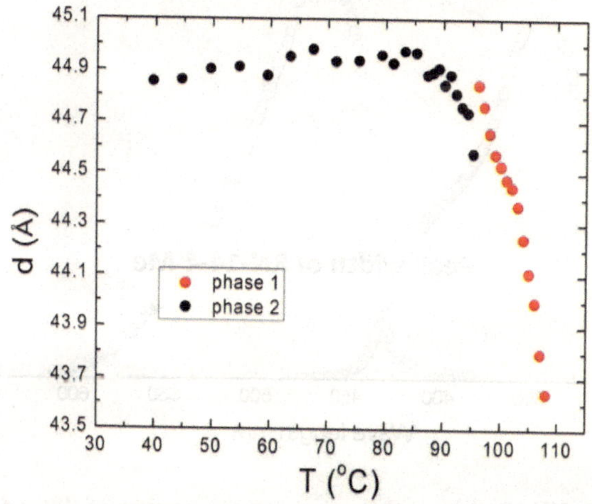

Figure: 6.23a: Variation on layer spacing with temperature

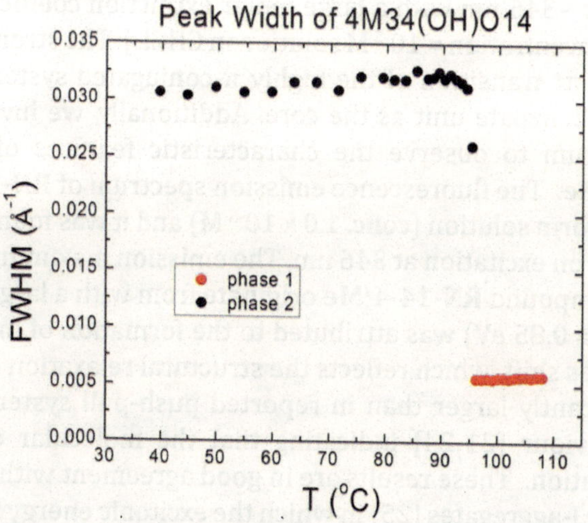

Figure: 6.23b: Variation of peak width with temperature

6.2.4: UV-Visible and fluorescence studies:

The UV absorption and fluorescence spectroscopic properties of compound **RN-14-4/Me** in solution were studied to obtain information regarding absorption and emission maxima and the Stokes shift of fluorescence. The UV-Visible and Fluorescence spectra of a representative compound **RN-10-4/Me** are shown in **Figure6.24**

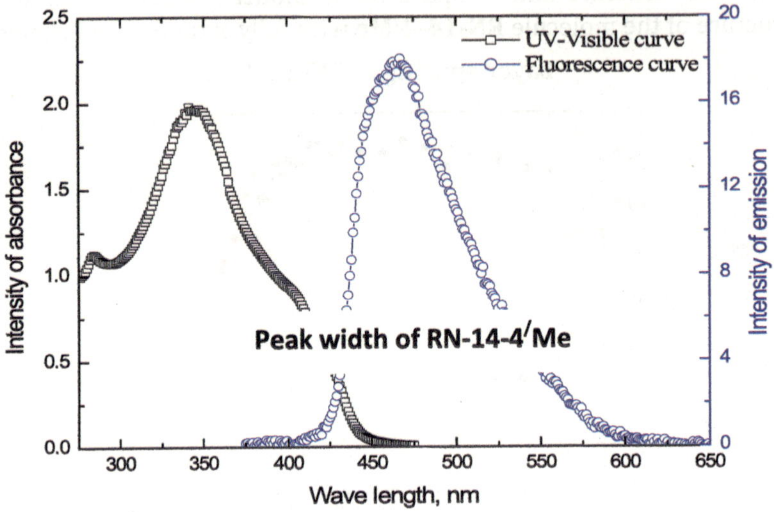

Figure 6.24: UV-Visible and fluorescence spectra of **RN-14-4/Me**

The UV-visible spectra in these homologous compounds exhibited strong absorption peaks at ~346 nm with a large molar extinction coefficient (3.59 eV for **RN-10-4/Me**, concentration = 10^{-5} M solution in $CHCl_3$). The strong absorption band reflects the π-π* transition of the highly π-conjugated system having the substituted phenyl benzoate unit as the core. Additionally, we investigated the fluorescence spectrum to observe the characteristic features of the excited states of **RN-14-4/Me**. The fluorescence emission spectrum of **RN-14-4/Me** was examined in chloroform solution (conc. 1.0 x 10^{-5} M) and it was found to exhibit a strong fluorescence on excitation at 346 nm. The emission maximum (λ_{em}) at 453 nm (2.74 eV) for compound **RN-14-4/Me** originate from with a large Stokes shift (λ_S) of 107 nm (ΔE = 0.85 eV) was attributed to the formation of intermolecular excimers. This Stokes shift, which reflects the structural relaxation of the excited molecule, is significantly larger than in reported push-pull systems exhibiting liquid crystal behaviour [21-24] indicating that the molecular conformation changes upon excitation. These results are in good agreement with the reported results of molecular J-aggregates [25] in which the excitonic energy is delocalized as a result of intermolecular coupling within the head-to-tail arrangement of the molecules in the solution.

6.4. Summary:

In this chapter we studied the mesophase behaviour of bent-shaped four-ring compounds which are substituted by methyl group at the central core in different position. By a systematic synthetic design we were able to compare the mesophase behaviour of analogous compounds which possess the different substituents at different position of the central core and varying the length of the terminal chains.

Summarising the results it is obvious that the influence of the substituents at the central core is dominant. This is not surprising since substituents attached at the central phenyl ring near to the connecting groups can strongly influence the bending angle between the legs and therefore the conformation of the whole molecule. Optimal bending angles have been found to be between 110–140u.

6.5. Conclusion:

In this book the influence of substitution at the bent aromatic core was investigated with respect to its effects upon the structures of the mesophases formed by 4-ring bent-core molecules. Mesomorphic behaviour of the compounds were confirmed by differencial scanning calorimetry, polarising optical microscopy and X-ray diffraction studies. All the compounds were found to be liquid crystalline and exhibited a wide variety of smectic and banana polymorphism. Phase behaviour of the compounds are presented in **Table 6.5**. It was found that, formation of different mesophases is not solely determined by the types of substituent but also by their position. Substitution by methyl, chloro and fluoro group on the four ring system produced liquid crystals with a wide variety of mesophases.

Table 6.5: Structure - mesomorphism relationship

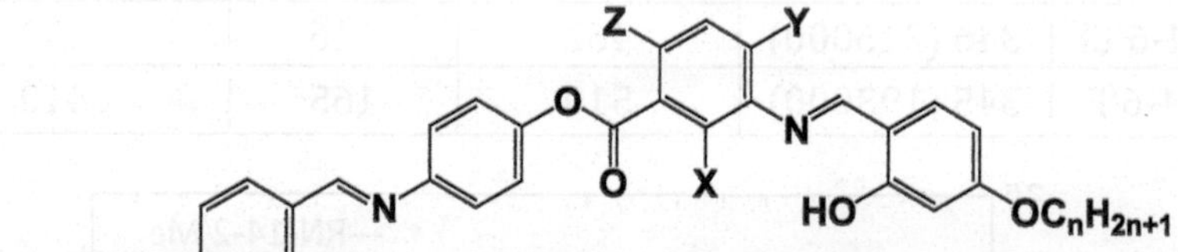

Compound	n=	X	Y	Z	Mesophases	Chapter
RN-n	9 - 16,18	H	H	H	$B1_{revTilted}$	4
RN -n-2/Me	6,8	CH_3	H	H	N*, SmA*, B2	6
	10,12,14,16,18	CH_3	H	H	SmC, $SmCP_A$	6
RN -n-4/Me	14	H	CH_3	H	SmA, B2, SmCP*(DC)	6
RN -n-4/Cl	10,12	H	Cl	H	B1	5
	14,16	H	Cl	H	B7	5
RN -n-6/Cl	14	H	H	Cl	$B1_{revTilted}$	5
RN -n-6/F	9 - 14,16	H	H	F	Col_r	5

To study the effect of substituent on absorption and emission spectra, one compound from each series of same chain length (n = 14) are compared and presented in **Figure 6.26.** The absorption spectra of all the compounds show absorption maxima (λ_{max}) at (344-346) nm and a low intense peak at around 410 nm. The more intense high energy band is associated with π- π* transition; the lower intensity band corresponds to a transition involving non-bonding orbital localized on the nitrogen atom, and leads to direct population of n-π*excited states. The fluorescence spectra of the compoundsare displayed in Figure 5.29b. It is observed that none of these spectra have multiple peaks, which indicates that the emission occurs from the lowest excited state with the largest oscillator strength. The peak positions of the absorption and fluorescence spectra are summarized in **Table 5.5.** All of the compounds show large Stokes shift ranging from 107 nm for **RN-14-2/Me** to 165 nm for **RN-14-6/F.** The fluorescence Stokes shift of **RN-14-6/F** is much higher as compare to other compounds. This means that the energy gap between the ground and lowest excited states decreases monotonically with fluoro substitution.

Table 6.6: Absorption and emission characteristic of substituted 4-ring systems (RN-14-n/X), values in the brackets are molar extinction coefficients (ε) of absorption.

Compound	Absorption Wave length (λ_{Max}) nm	Emission Wave length (λ_{Em}) nm	Stokes shift (Δλ) nm	Cut off wavelength ($\lambda_{cut\ off}$) nm
RN-14-2/Me	344 (194000)	451	107	420
RN -14-4/Me	346 (196000)	455	109	450
RN -14-4/Cl	345 (140000)	454	108	445
RN -14-6/Cl	346 (216000)	462	116	450
RN -14-6/F	345 (193000)	510	165	410

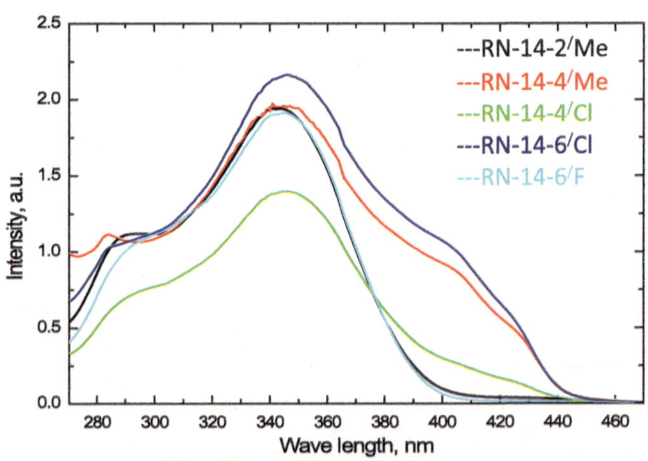

a) UV-Visible spectra

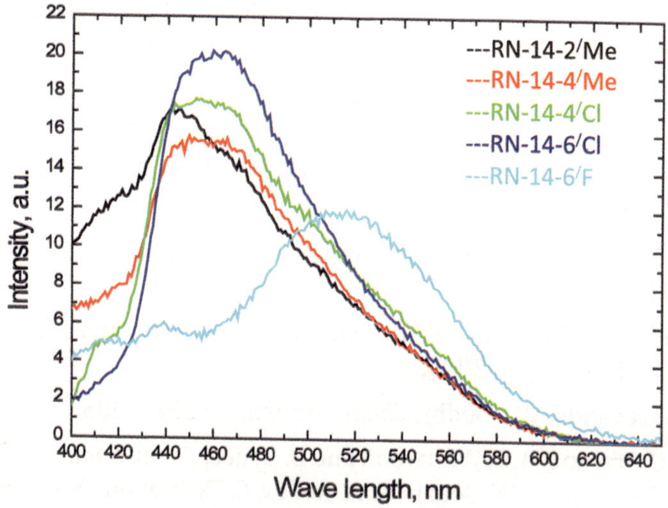

b) Fluorescence spectra

Figure 6.27: UV-Visible and fluorescence spectra of substituted 4 ring compounds

References:

1. W. Weissflog, in *Handbook of Liquid Crystals*, ed. D. Demus , J. Goodby, G. W. Gray, H.-W. Spies and V. Vill, Wiley-VCH, Weinheim, 1998, **vol. 2B**, 835.
2. W. Weissflog, C. Lischka, I. Benne´, T. Scharf, G. Pelzl, S. Diele and H. Kruth, *Proc. SPIE: Int. Soc. Opt. Eng.*, **14**, 3319, 1998.
3. S. Diele, S. Grande, H. Kruth, C. Lischka, G. Pelzl, W. Weissflog and I. Wirth, *Ferroelectrics*, **212**, 169, 1998.
4. G. Pelzl, S. Diele, A. Jakli, C. Lischka, I. Wirth and W. Weissflog, *Liq. Cryst.*, **26**, 135, 1999.
5. T. Sekine, T. Niori, M. Sone, J. Watanabe, S. W. Choi, Y. Takanishi and H. Takezoe, *Jpn. J. Appl. Phys.*, **36**, 6455, 1997,.
6. G. Pelzl, S. Diele, S. Grande, A. Jakli, C. Lischka, H. Kresse, H. Schmalfuss, I. Wirth and W. Weissflog, *Liq. Cryst.*, **26**, 401, 1999.
7. W. Weissflog, C. Lischka, S. Diele, G. Pelzl, I. Wirth, S. Grande, H. Kresse, H. Schmalfuss, H. Hartung and A. Stettler, *Mol. Cryst. Liq. Cryst.*, **333**, 203, 1999.
8. H. Hartung, A. Stettler and W. Weissflog, *J. Mol. Struct.*, **31**, 526, 2000.
9. J. Matraszek, J.Mieczkowski, J. Szydlowska and E. Gorecka, *Liq. Cryst.*, **27**, 429, 2000.
10. W. Weissflog, L. Kovalenko, I. Wirth, S. Diele, G. Pelzl, H. Schmalfuss and H. Kresse, *Liq. Cryst.*, **27**, 677, 2000.

11. I. Wirth, S. Diele, A. Eremin, G. Pelzl, S. Grande, L. Kovalenko, N. Pancenko and W. Weissflog, *J. Mater. Chem.*, **11**, 1642, 2001.
12. 12. L. Kovalenko, W. Weissflog, S. Grande, S. Diele, G. Pelzl and I. Wirth, *Liq. Cryst.*, **27**, 683, 2000.
13. G. Pelzl, A. Eremin, S.Diele, H. Kresse, W. Weissflog, *J. Mater. Chem.*, **12**, 2591, 2002.
14. T. Niori, J. Yamamoto, H. Yokoyama, *Mol. Cryst. Liq. Cryst.*, **409**, 475-482, 2004.
15. V. Gortz, C. Southern, N. W. Roberts, H. F. Gleeson and J. W. Goodby, *Soft Mater*, **5**, 463-471, 2009.
16. V. Gortz and J. W. Goodby, *Chem. Commun.*, 3262, 2005.
17. (a) C. Prasang, A. C. Whitwood and D. Bruce, *Chem. Commun.*, 2137, 2008; (b) P. S. Salter, P. W. Benzie, R. A. Reddy, C. Tschierske, S. J. Elston and E. P. Raynes, *Phys. Rev. E*, **80**, 031701, 2009.
18. T. P. Rieker, N. A. Clark, G. S. Smith, D. S. Parma, E. B. Sirota, C. R. Safinya, *Phys. Rev. Lett.*, **59**, 2658, 1987.
19. N. A. Clark, T. P. Rieker, *Phys. Rev. A*, **37**, 1053, 1988.
20. H. Gallardo, R. Cristiano, A. A. Vieira, R. A. W. Neves Filho, R. M. Srivastava and I. H. Bechtold, *Liq. Cryst.*, **35**, 57–63, 2008.
21. A. A. Vieira, R. Cristiano, A. J. Bortoluzzi and H. Gallardo, *J. Mol. Struct.*, 875, 364–371, 2008.
22. R. M. Srivastava, R. A. W. Neves Filho, R. Schneider, A. A. Vieira and H. Gallardo, *Liq. Cryst.*, **35**, 737–742, 2008.
23. R. Cristiano, F. Ely and H. Gallardo, *Liq.Cryst.*, **32**, 15–25, 2005; (e) A. A. Vieira, R. Cristiano, F. Ely and H. Gallardo, *Liq. Cryst.*, **33**, 381–390, 2006.
24. F. Camerel, L. Bonardi, M. Schmutz and R. Ziessel, *J. Am. Chem. Soc.*, **128**, 4548–49, 2006.
25. Y. Takanishi, H. Takezoe, A. Fukuda, H. Komurab and J. Watanabe, *J. Mater. Chem.*, **2(1)**, 71-72, 1992.